College of Astrology

Jaipal Singh Datta

ISBN:146807489X
ISBN-13:9781468074895

DEDICATION

This book is dedicated to my Astrology Guru Mr. H.R.Seshadiri Iyer of Bangalore, India. I never believed in Astrology. He taught me this Vedic Science. Astrology is a scientific subject. He predicted that I shall learn this science from him and spread this knowledge around the World.

Note

Due to many errors I have tried my best to correct my following books in January 2016. Please inform your address for free shipment of these books if you have purchased before January 2016 from Amazon.com

1.0   Astrology Made Easy

2.0   Astrology

3.0   Astrology Lessons

Astrology Lessons

CONTENTS

	Acknowledgments		i
1	Astrology Lesson No. 1	Elements of Astrology	4
2	Astrology Lesson No.2	Panchanga / Ephemeres	16
3	Astrology Lesson No.3	Planets , Rasis, Bhavas, Stars	24
4	Astrology Lesson No.4	Ayanamsha, Ephemeres, Tatwa,	28
5	Astrology Lesson No.5		38

Erection of birth Chart, Panchanga, Balance of birth, Periods , Sub Periods,

6	Astrology Lesson No. 6	Fixing Cuspal Chart, Bhava Chart	46
7	Astrology Lesson No. 7	Shadbala,	54
8	Astrology Lesson No.8	Different Yoga	58
9	Astrology Lesson No.9	Female Horoscopy	64
10	Astrology Lesson No.10	Results of Bhava in predictions	67
11	Astrology Lesson No. 11	Bhava Phalam Continues	84
12	Astrology Lesson No.12		88

Hora , Thithi, Yoga , Karna, Graha Samyam, Rectification, Family, Travels, Ayanamsha, Yogas,

13	Astrology Lesson No. 13	Thithi and Yoga Astrology	104
14	Astrology Lesson No. 14	Bhava Chart and Rasi Chart	110
15	Astrology Lesson No. 15	How to erect Division Charts	118
16	Astrology Lesson No.16	Magnitude of effects	124
17	Astrology Lesson No. 17	Division Charts	130
18	Astrology Lesson No. 18	Dasha Bhukti Readings	138
19	Astrology Lesson No. 19	Gochara	144
20	Astrology Lesson No. 20	Rahu Kalam , Thithi, Weekday, Stars , Yoga , Kalasarpa Yoga, Kuja dosha	150
21	Astrology Lesson No. 21	Marriage Alliances	160

Astrology Lessons

ACKNOWLEDGMENTS

I am thankful to Padma Shree, Brigadier (Retd.) Dr. Kapil Mohan V.S.M., PhD and Smt. Pushpa Mohan for their encouragement to write knowledge about this novel subject for the coming generations. I am thankful to my parents, elders, Mohyals, and my global family. I am thankful to my wife, my children for their love and affection and support to me.

We all are puppets in the hands of destiny. We are governed by law of Nature. I acknowledge that I am nothing. I am only acting as per desires of planets and stars controlled by Elephant head Lord named as Ganesha, who is told to be son of Lord Shiva and Parvathi.
OM GAN GANPATHAYE NAMAH

LESSON NO.1

Prayer

Before you commence to read these lessons please offer your prayers first to God Ganapathy, then to your Ishta Devatha (Family Deity) and finally to your Guru (Preceptor).

Books to be possessed

Our Publication "New Techniques of Prediction" in two parts (hereafter referred to in our lessons as merely our Part I or II) Lahiri Ephemeris and Tables of Houses or Raphael's Ephemeris and Tables of Houses and for those who need a Panchanga a standard Drigganitha Panchanga as in the South of Kumbakonam Mutt Srouthi Panchangam or any other one which tallies with the positions cited in Lahiri's Ephemeris. We dissuade you from reading any other book on this subject as they will only confuse and mislead you from the real track. If in the course of our lessons we quote any authority you may not go in for that entire book referred by us as being entirely genuine. For, it is only that particular portion that we value most and no more. As our publications and teachings are deep study and practice of all the books available and by research work made so far, if you merely follow our lessons with our above stated treatises you will have acquired the essence of all the other books that are available in the market and even something more than it as we impart some secrets of Nadi Astrology we have learnt at the feet of many Gurus. For abbreviations of planets see our Part I Page I.

II. Method of our condensed Pundit Course

3

As we have already discussed at length in our texts cited above most of the subjects at length we will in these lessons elaborately discuss on points which are briefly stated in our books and those that are not at all mentioned in them coupled with the latest research results. In other cases of elementary nature we merely refer to the chapters and pages of our books touching the most important items only and adding some special notes where necessary. As lessons are sent on each payment the students will go through them and if they feel any doubt on any point they may be referred to us allowing wide margins of spaces against the questions for reply. All such replies will be sent within a week or two.

III. If for any reason the lessons are not received within a week after payment the same may be brought to our notice.

These lessons are sent on the solemn understanding that they should be made use of by the contributor only. Please make it a point to keep on monthly regular remittances in time to facilitate timely and regular dispatch of lessons as they will be automatically held back for want of timely receipt of the fees. With these we bless you and proceed further.

IV. What is Astrology?

Astrology is a branch of Veda (Vedanga) that has come down to us from God Ishwar with the aid of which we can forecast all the past, present and future events considering the positions of planets and Lagna at birth, query or any moment. It has three broad divisions namely Siddhantha, Samhita and Hora. Siddhantha deals with the astronomical side of Astrology, Samhita with the World events in general and Hora with

Horoscopy of human and animate objects on this earth.   In golden days when mathematical ready-made tables were not available they used to work out from the fundamental elements troubling themselves a lot to arrive at an astronomical calculation may be, even so not quite correct for very many reasons such as the various systems of existing Siddhantha. The different reckonings of Ayanamsha and want of knowledge of fractions and decimals which are only of recent origin were also setbacks to them. Thus to cast a Horoscope rightly or wrongly it would take them a lot of time and labor. Many a time it would be "Love's Labor Lost". To understand any Siddhantha a student should devote a couple of years as there are various Siddhantha.

Now-a-days there are ready-made Ephemeris giving day-to-day positions of planets with all the other characteristics such as Latitude, Declination etc., and with the aid of these Ephemeris one can cast a horoscope correctly in a few minutes. You must follow the correct Ephemeris. For this read our advice above. Thus the most difficult and intricate part of Siddhantha is solved in a minute. We have thus crossed a handicap of two years' study.

Samhita deals with the affairs of the world events such as seasonal effects, world events, war, epidemics, effects of eclipses, market prices, national events, weather forecasts, geological, survey etc. We are at present concerned with Horary Astrology dealing with horoscopy and query. In these lessons we will discuss Horary Astrology as foremost.

V.   In Horary Astrology what are all the Fundamentals to know?

We have to know that there are:

1. A Zodiac of 12 Rasis   (Mesha to Meena)
2. Nine planets   (Sun to Ketu)
3. 27 Stars   (Ashwini to Revati)
4. 60 years   (Prabhava to Akshaya)
5. 6 Ritu   (Vasanta to Sishira)
6. 12 Solar months   (Mesha to Meena)
7. 12 Lunar Months   (Chaitra to Phalguna)
8. 2 Paksha   (Shukla and Krishna)
9. 7 Week days   (Sunday to Saturday)
10. 30 Thithi   (Shukla Padyami to Amavasya)
11. 27 Yogas   (Vishakamba to Vaidriti)
12. 9 Karana   (Bhava to Nagava)

For further details of the above elements please read Astrology made easy.

Special Notes

A. Hindu or astrological week day is from sunrise to sunrise and not from midnight to midnight as per English reckoning.
B. By 30 Thithi are meant 1 to 14 in Shukla Paksha – Padyami to Chaturdashi, 15th being Purnima, 16 to 29 is again Krishna Paksha – Padyami to Chaturdashi, 30th being Amavasya.

VI.   Next you must know the relationship of planets with reference to Rasis

    1. Ownership

We will disclose a beautiful point here as to why and how specific houses are allotted to planets in the way it is stated. You will see from the following logical arguments that there is a definite principle underlying all theories.   First of all we shall discuss the theory of ownership of all planets. You must first of all know that planets have their own orbits of motion in the Heavens. By orbit is meant the path in which they move. They are fixed. Each planet moves in its own orbit at different heights from the earth.   The following are the order of orbits in increasing distant from earth - Moon – Sun – Mercury (Budha) – Venus (Sukra) – Mars (Mangal / Kuja) – Jupiter (Guru) – Saturn (Sani).   Thus Moon is nearest to earth and Saturn is the farthest. It is due to the nearness of Moon that much importance is attached to Moon as having great influence on living beings and vegetarians on earth.   Their speed of movements are likewise. Moon is the fastest moving planet while Sani is the slowest.

    Except Sun and Moon the rest have ownership over two houses.   In fact Sun and Moon are luminaries while the rest are planets that get illumination from the luminaries.

    Place the luminary Moon in Kataka and successive planets according to their orbits one in each Rasi in anti-clockwise order starting from Mithuna which is next to Kataka. Similarly place the Sun in Simha and the rest in order from Kanya onwards in clock-wise order.   The houses in which they will be thus placed will be the houses owned by those planets. Then the order of their distribution will be as follows:-

Budha – Sukra – Kuja – Guru and Sani. It is for this reason that the 6 Rasi from Kataka to Kumbha in anti-clockwise order are called Chandra Rasis or Moon Street while the other six are called Surya Rasis or Sun Street. By so working we find.

Moon owns Kataka	&		Sun owns Simha
Budha	owns		Mithuna & Kanya
Sukra	owns		Vrishbha & Thula
Kuja	owns		Mesha & Vrischika
Guru	owns		Meena & Dhanush
Sani	owns		Kumbha & Makara

P.S. Note the above order of Rasis on either side of Kataka and   Simha.

Special Notes

(a) The above order of orbits of planets are also helpful to find out the fast and slow moving planets (Sheeghra gathi and Manda gathi) and this is useful when we deal with aspects (applying and separating). For example between Budha and Saturn, Budha is a faster moving planet. Between Moon and Kuja Moon is faster and so on.

(b) A planet in own house is supposed to be cool and happy.

(c) Rahu owns Kumbha and Ketu owns Mesha (specials).

8

2. *Exaltation and Depression signs with maximum Degrees*

Planets	Exaltation Rasi	Depression Rasi	Max Degrees
Sun	Mesha (Aries)	Thula (Libra)	10
Moon	Vrishbha	Vrischika	3
Mars	Makara	Kataka	28
Mercury	Kanya	Meena	15
Jupiter	Kataka	Makara	5
Venus	Meena	Kanya	27
Saturn	Thula	Mesha	20
Rahu	Vrischika	Vrishbha	Not Fixed
Ketu	Vrischika	Vrishbha	Not Fixed

Special Notes

(a) At exact opposite point of exaltation (180^0) is the planet's utmost debilitation point. A planet is said to be in exaltation when it sends its rays to earth directly at right angles without forming any angular rays. You know that direct rays are very powerful as you experience the mid-day rays of the sun to be very severe. Reverse is the case of debilitation.

(b) It is a misconceived notion that an exalted planet will always do good. The truth is that he will be endowed with full strength to do good or bad according to his trait

9

in the horoscope. It only speaks of the strength and not of the nature. This will be discussed in detail in our future lessons.

3. Moolatrikona House

The following are the Moolatrikona houses of Sun to Saturn in order. Simha – Vrishbha – Mesha – Kanya – Dhanush – Thula – Kumbha respectively i.e. Sun in Simha, Moon in Vrishbha, Mars in Mesha, Mercury in Kanya, Jupiter in Dhanush, Venus in Thula, and Saturn in Kumbha.

Special Notes

(a) Of the two houses owned by a planet one is Moolatrikona.
(b) A planet in Moolatrikona is said to possess the effect as if in his own house but of a slightly higher order.

4. Friendly, Inimical and Neutral Planets

Planets	Friends	Enemies	Neutrals
Sun (Rv)	Ch, Kj, Gr	Sk, Sn	Bd
Moon (Ch)	Rv, Bd	None	Kj, Gr, Sk, Sn
Kuja (Kj)	Ch, Gr, Rv	Bd	Sn, Sk
Budha (Bd)	Rv, Sk	Ch	Kj, Gr, Sn
Guru (Gr)	Rv, Ch, Kj	Bd, Sk	Sn
Sukra (Sk)	Bd, Sn	Rv, Ch	Kj, Gr

Sani (Sn)   Bd, Sk   Rv, Ch, Kj   Gr

Special Notes

(a) For abbreviation see part I Page I.
(b) A planet in friendly sign creates friendly atmosphere, one in enemy's sign give inimical effects while that in neutral sign remains neutral in character.
(c) Apart from the above natural friendship the planets that are mutually in 2-12, 3-11, and 4-10 signs from each other are said to be temporary friends, but we may not attach much weight to it when we deal with birth charts. They may be useful in Prasna or query time.
(d) Signs owned by friendly etc., planets will be likewise friendly etc., signs. This will be useful when predictions are based on relative characters of Rasis as in Jaimini etc., systems where Rasi Dashas etc., are considered.

VII. Next we must know the condition of planets

1.   A planet coming near sun is said to be combust or Asta or Set. The nearer he is to Sun the greater the combustive effect. Different ranges of combustive effect are ascribed to the various planets but for our purposes one within 3^0 range from Sun may be taken as severely combust.

One in the exact degree of Sun is totally lost and becomes helpless to give any effect and tends to give bad effects. Even here one difference has to be noted. If the combustion is applying it will be virulent while separating it will be milder and fading of the bad effects. For periods of setting or combustion

11

see the Ephemeris or Panchanga. Thus while reading the effects of combustion you must take special note of the distance as well the nature of the aspect applying or separating.

Special Notes

    (a) In the case of Sukra and Sani only half their power is said to be reduced by combustion.

    (b) If Sun is Birth Yogi (which will be taught later) or Duplicate Yogi then the planet in combustion will give Yoga (prosperity) instead of adversity.

A Retrograde Planet

2.   When a planet appears to move in anti-clockwise order we call it retrograde. Sun and Moon have no retrogression while Rahu and Ketu are natural Retrogrades. We say it appears to move backwards. Please note this. In the heavens (orbits) they never have a backward motion. For those on earth it appears to be so. Such a phenomenon is seen and caused when a planet comes very near the earth in its orbit of Revolution.

    In our future lessons on Shadbala we will deal with one type of strength called 'Kalabala'. As per this Sun gets Kalabala in Uttarayana. Moon in Shukla Paksha and rest when they are retrograde. So it follows that a retrograde planet is powerful only so far the strength is concerned and should not by itself be concluded that it will always give very good effects. The good or bad has to be decided by other factors while only strength has to be estimated by his retrogression.   Fuller detail may be had on this topic in our future lessons.   It only speaks of the strength and not nature.   So follows that a retrograde planet is powerful whenever he is in exaltation, debilitation or any sign. There are some who babble that retrogression in exalted sign gives Neecha effect etc.   All the myths may be shelved to the corner and the most sensible and cogent theories narrated by Daivaganas only adopted.

Read this lesson carefully and understand the same properly before you read the next lesson No.2.

Names of English to Hindi

Planet	English to Hindi	Short
Sun	Ravi	Rv /RV
Moon	Chander	Ch /CH
Mars	Mangal / Kuja	Kj / KJ
Mercury	Budha / Buddha	Bd /BD
Jupiter	Guru	Gr /GR
Venus	Sukra	Sk / SK
Saturn	Shani / Sani	Sn / SN

May the Almighty grace you with intellect and insight.

Rashi Sign in English to Hindi

Aries	Mesha	Taurus	Vrishbha
Gemini	Mithun	Cancer	Kataka
Leo	Simha	Virgo	Kanya
Libra	Tula	Scorpio	Vrischika
Sagittarius	Dhanush	Capricorn	Makar
Aquarius	Kumbha	Pisces	Meena

Read this lesson carefully and understand the same properly. May the Almighty grace you with intellect and insight.

2 LESSON NO. 2

In lesson 1, you were required to get yourself well acquainted with the several Astrological Elements described in Chapter 1, Part 1 of our Text (N.T.P.) or Astrology or Astrology made easy. Of these you must specially know more about Thithi and Yoga as they are of Paramount importance in shaping the destinies of persons. So far, no Astrologer has been using them as he is not at all aware of its theory and much less its use. You will be the first to know these valuable secrets in this lesson. This lesson is of great importance.

As stated in Lesson 1, there are 30 Thithis, 27 Yogas and 11 Karanas. For those born in a Thithi some Rasis get worsened called Zero-Rasis (vide Table on pages 9-9 Part 1). The effects of these will be explained in later lessons.

These three elements may be known either from a Panchanga or an Ephemeris.

Panchanga Method

Against the day of Birth the names of Thithi, Yoga and Karana are stated with their durations in Ghati or hours. Your business is only to find out the corresponding ones ruling exactly at the Birth Time. Here you get merely the name of the Yoga. To locate it in the chart you have to find the equivalent star of the Yoga (vide PP 10-11, Pt. 1) and then fix the Yoga point at the place where that star is situated. For starry positions & their equivalent Degree positions see below. The Lord of this Star (Udu Dasha Lord) becomes the Birth-Yogi who plays a very important role in predictions.

How to Locate these from Ephemeris?

On pages 38 to 41 of Lahiri's Annual Ephemeris the durations of the Days' Thithi, Yoga and Star are stated in Indian Standard Time (I.S.T.). For the Birth time of that day find out the corresponding elements. In the absence of such Tables, you must do the original work as follows:-

To Find Thithi

From the longitude of the Moon subtract the longitude of Sun and divide the balance by 12 Thithi corresponds to the Quotient plus 1 (counted from Shukla Padyami). For easy understanding convert the Rasis and Degrees of the positions of planets into total degrees and minutes (Counting from 0^0 Mesha). If Moon's Degree is less than that of Sun add 360^0 (Total of 12 Rasis) to that of Moon and then subtract.

For example, let us suppose a birth at 5.30 a.m. on 13th August, 1963. Sun's position at that time is 3 Rasis 26^0 -11', i.e. 3-26-11 or $116^0 - 11'$ and that of Moon (1-5-54) is $35^0 - 54'$. As Moon's total longitude is less than that of Sun and we cannot subtract as it is, we add 360^0 to Moon and we get:-

Moon's position at Birth Time = 395^0 -54'

Sun's position at birth time = 116^0-11'

Subtracting we get = 279^0 -43'

Dividing this by 12 we get 23 for quotient and 3^0-43' as Remainder.

So the Thithi is 23 + 1 = 24th

Counting from Shukla Padyami, it is Krishna Paksha Navami – As it is Navami, the zero Rasis are Simha and Vrischika (Pp 8.9 Pt. I).

There are 27 stars and 27 Yogas. Vayu Devta are 49. Rudra Devta are 11 and Vasu are 8. So total are 68.

How to Find Karana?

A Thithi has two Karanas – the first half has one and the second half has another. As the Span of a Thithi is 12^0, its half is 6^0. So the span of a Karana is 6^0. If the Remainder got above be less than 6^0 its former Karana is to be reckoned and if more than 6 the latter one is to be taken. The distribution of Karanas varies according to Paksha (see table on Pp. 16 Pt. (I). In the above example, the balance is less than 6^0 so it is the first half Karana of Krishna Paksha Navami. As per the said table, it is *Thaithula Karana*. So you can locate the Karana.

Why should you divide by 12 to get a Thithi?

You should know that all the 30 Thithis do by equal distribution occupy the 12 Rasis, i.e. 360^0 . So the span of each Thithi is 360/30 = 12^0 . Hence to find the Thithi we should know how many such 12. Hence to find the Thithi we should know how many such 12 degree have elapsed from Sun to Moon (as these two cause the Thithis).

Add to the sum of the longitudes of Sun and Moon a constant measure of 3 Rasis 3 Degree 20 minutes or $93^0 - 20'$. You get a point called Yoga point. The Star represented by that position becomes Yoga Star and the lord of that Star (Udu Dasha Lord) which you will know later (vide Pp. 18 Pt. I) becomes the Birth Yogi who plays a very important role in Horoscopy and Horary Astrology also.

In the above example:-

Sun	$= 116^0 -11$
Moon	$= 395^0 -54$
Constant Factor	$= \underline{\ 93^0 -20}$
Total	$= 605^0 - 25$

(P.S.) if it exceeds 360^0 substract

$\underline{360^0 \text{ -00'}}$ (one Zodiac Measure)

$= 245^0$-25 this is Rasis, etc., will be 8-5-25.

i.e. 5 Degree -25' of Dhanush and this corresponds to Moola 2nd Pada. The lord of Moola star is Ketu. So Ketu is the Birth Yogi.

By Ephemeris method you get the Yoga Star direct. If you want to know the name of the Yoga find its equivalent from the table on Pp. 10-11, Pt. I.

Why should the constant factor of 3-3-20 be added in all cases is explained on Pp.11-1. As the starting point of Yoga is Vishkambha and its equivalent is Pushyami start located at 3-3-20 we add this in all cases.

Thus you get at a stretch the Thithi, etc., from the Ephemeris method.

Below is stated the situation of stars in the Rasis and the longitudinal (Degree etc.) positions of these stars and finally the Lords of Stars. How are we to know these factors?

Distribution of Stars in Rasis

You have already read in Lesson 1, that there are 27 Stars running in a particular order from Ashwini to Revati. All these 27 stars are distributed equally and in consecutive order in the 12 Rasis of the Zodiac. Thus by Rule of three we get 2 and 1/4 Stars located in each Rasi.   For some convenience each star is divided into 4 equal parts called *Padas*. Thus a Rasi contains 2and 1/4 Stars or 9 Padas. Thus the entire Zodiac has 108 Padas (12 X 9 = 108) which number is given much importance in the chanting (Astothharam). Next the order:-

Starting from 0^0 of Mesha with Ashwini 1st part, we have in Mesha-Ashwini 4 parts, Bharani 4 parts and Kritika 1 Part totaling 2+1/4 stars or 9 Padas.   Next Vrishbha Rasi starts with Kritika 2, 3 and 4th Padas, Rohini 4 parts and Mrigsira 1st and 2nd parts.   Running further in the same order, lastly in Meena Rasi, 4th part of Poorva Bhadra, all the 4 parts of Uttara Bhadra and all 4 parts of Revati are situated. Thus you have to note the distribution of Stars. These are fixed starry positions.

How to apportion longitudinal spans to these stars?

You must know that the Longitudinal Span of a Rasi is 30 Degrees. Distributing these 30^0 to 9 Padas of a Rasi, we get $3^0 - 20'$ as the longitudinal span of a Star Pada. Above is stated the order of the

Stars in Rasis. Accordingly allot their Degree positions. For clearer understanding see the Table Below:-

Rasis	Stars	Span in Degree & Minute
1. MESH	Ashwini (F)	0 to 13-20
	Bharani (F)	13 -20 to 26-40
	Kritika (I)	26-40 to 30-00
2. VRISHABHA	Kritika (2.3.4.)	0 to 10-00
	Rohini (F)	10 to 23-20
	Mrigsira (1.2)	23-20 to 30-00
3. MITHUNA	Mrigsira (3.4)	0 to 6-40
	Aridra (F)	6-40 to 20-00
	Punarvasu (1.2.3)	20-00 to 30-00
4. KATAKA	Punarvasu (4)	0 to 3-20
	Pushyami (F)	3-20 to 16-40
	Ashlesha (F)	16-40 to 30-00
5. SIMHA	Makha (F)	0 to 13-20
	Pubba (F)	13-20 to 26-40
	Uttara (I)	26-40 to 30-00
6. KANYA	Uttara (2.3.4)	0 to 10.00
	Hasta (F)	10-00 to 23-20

20

	Chitta (1.2)	23-20 to 30-00
7. THULA	Chitta (3.4)	0 to 6-40
	Swati (F)	6-40 to 20-00
	Vishakha (1.2.3)	20-00 to 30-00
8. VRISCHIKA	Vishakha (4)	0 to 3-20
	Anuradha (F)	3-20 to 16-40
	Jyeshta (F)	16-40 to 30-00
9. DHANUS	Moola (F)	0 to 13-20
	Poorvashada (F)	13-20to 26-40
	Uttarashada (1)	26-40 to 30-00
10. Makara	Uttarashada (2.3.4)	0 to 10-00
	Shravana (F)	10 to 23-20
	Dhanishta (1.2)	23-20 to 30-00
11. KUMBHA	Dhanishta (3.4)	0 to 6-40
	Shatabhisha (F)	6-40 to 20-00
	P. Bhadra (1.2.3)	20-00 to 30-00
12. MEENA	P. Bhadra (4)	0 to 3-20
	U. Bhadra (F)	3-20 to 16-40
	Revati	16-40 to 30-00

(P.S.) Unless you master these and have them at your fingers' ends, you will feel the difficulty to follow future lessons. So master these definitions.

Abbreviations used above

(F) Full star meaning that all its 4 Padas are there. 1.2.3.4 = stand for the particular Pada or Padas of that star.

Special notes

From the above you may observe some particular and peculiar distribution of these stars in Rasis. Some have all their 4 Padas in a Rasi, some have half and half in two Rasis; some have only one Pada in a Rasi and some others 3 Padas.

Again some commence a Rasi and some others end with a Rasi. Each variety has its own special effects which will be described in later lessons.

Who are the Lords of Stars?

They are no other than the Udu Dasha Lords which you will know later. At present read Pp. 18 Pt. I for Stars and Dasha Lords.

You have had enough and more food to digest in this lesson. Be prepared to go through the next lesson after fully digesting all the previous lessons.

LESSON NO.3

If you want the detailed meticulous readings you have to master the Karakatwas of Planets, Rasis, bhavas and stars. It is by summing up all these you can give the considered effect of an aspect in life as any one consideration only may not always give out the correct version. For example in determining the sex of the first born child the lord of the 5th house may be a male planet and his period may also run then and yet a female child be born for reason of it being situated in female star though it be in a male Rasi. So to give the final judgment in such cases find out which of the above factors is most strong. The strength of a planet you know is by its Shadbala to be known later. The strength of a Rasi is that of its lord and that of a star is likewise that of its stellar lord. In this way you have to assess the strongest of the above three i.e., planet, Rasi and star. This necessitates you to know all the natural characteristics.

Planets

While judging the planetary affects you may know that there are two sides of the question.

1. The natural characteristics (Karakatwas) and

2. Functional characteristics (Adhipathya).

During their periods they cause these effects and in addition they also cause the effects of the Bhavas, Rasis and stars in which they are placed.   For more details on this subject please wait for the ensuing lessons.

As detailed characteristics of planets, Rasis, bhavas and stars are fully discussed in Ch. 11 Pt. 1 Pages 27 to 81of our Text we do not repeat them here.   Please go through them with care and try to remember at least the most important ones that may be used often. The rest may be referred to when necessary.   Some of the most important ones are detailed below for guidance.

Under Rasis

1.   Chara, Sthira and Dwiswabhava Rasis to be noted.
2.   Odd and Even Rasis, Male and female Rasis
3.   Childless Rasis – Item 9 pp. 28 Pt.1.
4.   Nations represented by them.

Under Planets

1.   Note the special traits on page 39 and the general characteristics from page 39 to 63 pt. I.
2.   Some of the special characteristics not fully described in the text are as follows:-
(a)   The numerals allocated to Sun on to Sani in order are – 1 & 4, 2 & 7, 9, 5, 3, 6 and 8 respectively.
(b)   The alphabets governed by :-
Sun – Avarga (Letters AH to Aha first nine vowels)

Moon – Ya to Ksha (the last 10 consonants)

Kuja – Kavarga (Ka, kha, ga, gha, gna)

Budha – Tavarga (Ta, ttha, da, ddha, nna)

Guru – Thavarga (Tha, thha, tha, thah, Na)

Sukra – Chavarga (Cha, chha, ja, jha, ingna)

Sani – Pavarga (Pa, pha, ba, bha, ma)

c)   The Thithis governed by them are:-

Sun – Prathama & Ekadashi (1 – 11)

Moon – Dwithiya & Dwadashi (2 & 12)

Kuja – Shasti (6)

Budha – Saptami (7)

Guru – Tritiya, Ashtami & Thrayodashi (3-8-13)

Sukra – Chowthi, Navami & Chaturdashi (4-9-14)

Sani – Panchami, Dashami, Full Moon and New Moon days

(5-10-15-30)

3. Next of importance that are in daily use are the Deities, Grains and Gems represented by the planets.

Under Stars

1. Norte the sex of stars – pp.64.
2. Stellar pars of body – pp. 69
3. All the other characteristics from pp. 70 to 81

Though this appears to be a short lesson it involves wide range of subjects for study. So please study them carefully and master them before we send you the next lesson.

N.B. – Please note that the pages referred to in these lessons slightly differ in our subsequent Editions of our Texts.

LESSON NO.4

In this lesson we deal with the preliminary precautions to be noted before casting a Birth Chart. As per point in our text (Pt. 1-Ch. III) most of the astrologers proceed straight-away with the mere Rasi chart as given by the party taking the trouble to verify the accuracy of the same. In our experience we have found that in almost all cases this may not be correct. Some of our consultants who banked on the accuracy of their charts prepared by learned astrologer's estimation have returned with the conviction that they were not really correct. This difference is due to several reasons as reference Panchanga or Ephemeres, Ayanamsha, mode of reckoning birth time & finally with all these also being noted small errors arising in the mathematical calculations. Even all these strictly done the time of birth need rectification. Under our theory we have proved that even a minute's difference in birth time will-capsize the entire readings. So it is the onerous responsibility of an astrologer to rectify the birth time & cast the chart correctly following right method. This important act we have succeeded in striking and in this lesson we impart most valuable theory so that you may not falter in future as others do. If you do not strictly follow these instructions you are sure to go wrong and thus spoil reputation & thus bring discredit to the science & your preceptor. It is the habit of every so-called renowned and famous Astrologer force his own opinion on others without taking the trouble of practical verification. You will have to ignore all and simply follow the one we state & put the same for further test for your personal conviction.

What are those Preliminaries?

Astrology Lessons by Astro Guru Datta

1. Ayanamsha

First & foremost this is the most controversial subject talked of by all in their own way. There was a time say 285 A.D. (which is the only correct version that has stood test in all cases without exception) when the fixed first point of Mesha coincided with the Vernal Equinox-the movable point. Since this year the movable point moves at an average of 50.4 seconds per year. That got for the year is termed Ayanamsha. The controversy in this matter is regarding the year of the said coincidence of the two points. All the astronomers base their arguments on hypothetical data but none take the trouble of verifying their results on the practical side of horoscope for, usually an astronomer will not be good astrologer. We have applied all the different sets of Ayanamsha & finally are of candid opinion that one adopted by the Govt. of India on the Report of the Calendar Reform Committee in March, 1957 is the correct one & that is what is followed by Mr. Lahiri & Kumbakonam Mutt Panchangam, Turn a deaf year to all the other opinions & simply follow this.

2. Ephemeres to be followed

As stated on PP. 83 Pt. I of our text it is only a standard Drigganitha Panchangam or Raphael Ephemeres or Lahiri's Ephemeres with their Table of Houses that are authentic. So possess copies of any one set. For Lahiri's sets address of Smt. N. Lahiri, Astro Research Bureau, 57/6, Raja Dinendra Street, Calcutta-6.

Birth Time

As discussed at length on PP 85 Pt. I, take the time when the child inhales breath for the first time & this shall be only after the umbilical cord is cut. As it is very difficult to note the first breath the entire responsibility of detecting this rests on the ability of the

29

astrologer. To do this we should know the method of rectification of birth which we teach in this lesson. We adopt two systems – Tatwa Siddhantha & Phala Kundali or Division charts method. At present we teach you the system of Tatwa Theory.

Tatwa Siddhantha

As this is explained at length in pages 86 to 88 Pt. I we do not wish to repeat them here. So please go through them in detail & note the following special points.

Chief points to be noted therein are:

a) The Tatwa move in a particular order.

b) They have two cycles one Aroha & the other Avaroha, each cycle being of 1and 1/2 hrs. 90 Minutes duration.

c) Each Tatwa has its own duration.

d) Each Tatwa belongs to a sex.

e) Tatwa govern Week Days.

f) The Tatwa of the week day commences at sunrise on that day followed by the successive ones in their prescribed order observing the cycles of Aroha & Avaroha.

g) Sometimes the sex changes at the transition point from either Aroha to Avaroha or from   Avaroha to Aroha.

We clarify by an example below:

Tuesday

At sunrise on Tuesday Tejo Tatwa / Agni Tatwa or Fire Tatwa starts & rules for 18 minutes & then they move as follows:-

This is first Aroha Cycle

Thejo (Agni / Fire / Carbon)	18'
Vayu (Air / Oxygen)	24'
Akash (Energy)	30'
Prithwi (Earth / Nitrogen)	6' .
Appu (Water / Hydrogen)	<u>12'</u>
Total	<u>90'</u>

Then after the first 90' of Aroha cycle of Tatwa from sunrise the first Avaroha cycle starts with the following order:-

Appu	12'
Prithwi	6'
Akash	30'
Vayu	24'
Tejo	<u>18'</u>
Total	<u>90'</u>

There after Aroha & Avaroha cycles repeat in the above order & move still the next day's sunrise i.e., Wednesday & on Wednesday at sunrise the Tatwa of the day i.e., Prithwi starts.

How does this Tatwa theory help us in fixing up the birth time?

For the given birth time find out the Tatwa that rules & note its sex. If this tallies with the sex of the native then say the time of birth is under this first correction correct. If it shows the opposite sex (except at transition points) then adjust the nearest Tatwa which tallies with the sex of the native. Thus you get the first set of corrections. Sometimes due to the longer interval of a Tatwa period you may not be able to fix up to the minute. For further perfection you must resort to Phala Kundali or Division Charts. Wait for the lessons on Division Charts.

Special Notes

1. We have spoken of the change of sex at transition points & this we illustrate for clear conception. If one is born at exact 1½ hours after sunrise when it would be aroha Appu end point (transition point from aroha to Avaroha) instead of a female sometimes male may be born. Similarly at exact 3 hours after sunrise when there is Tejo Tatwa at transition instead of a male a female may be born. In either case there will be some sex peculiarity arising. The male may inherit the qualities of a woman while the woman the qualities of man.
2. We have also experienced that generally births take place either at the commencement or end of a Tatwa, or midpoint.

These will help you to fix up to a nearer point of birth time.

Do not be misled by anybody's version that there is no Avaroha & only aroha is in vogue. In support of our theory please read the shloka on PP.88 Pt. I. As its meaning is not given in the book we shall narrate its meaning here.

Tejas, Appu, Tejas, Prithwi, Akash, Appu, Vayu are governed by Sun onwards in the order of their week days i.e. Moon, Kuja, Budha, Guru, Sukra & Sani and Sun. Commencing from Prithwi Tatwa as ¼ Ghati (6'0 the other Tatwa i.e. Tejas, Vayu & Akash each go on increasing by ¼ Ghati i.e., Prithwi 1/4, Appu ½, Tejas- ¾, Vayu-1, & Akash 1¼. In the two Yama's (Yama-3¾ Ghati or 1½ hrs.) called a Prahara (3 hrs.) the former Yama is aroha & the latter Avaroha. It goes on repeating every Prahara (3 Hrs.). Female is born in Appu (Water / Hydrogen) & Vayu (Air / Oxygen) Tatwa & Male in Prithwi (Earth /Protein / Nitrogen), Akash (Energy) & Tejo (Agni /Fire / Carbon) Tatwa. Like this you must fix up the birth time & then Lagna.

On PP.90-91 Pt. I. We have shown the method of working out the Antharas Tatwa etc., to help further nearer. If you can follow it read there, otherwise stop at major Tatwa at present as finalization will have to be struck by method of 1) – charts.

The effects of persons born under several Tatwa on PP91. They invariably prove to be correct. Of them Tejo gives more correct & in it those born in aroha Tejas become notable personalities. Please digest these preliminaries well before the casting of chart in the next lesson. Please note that at this stage you have almost covered all basic of Astrology

LESSON NO.5

Erection of Birth Chart

To do this you must know the fixing up of planets. Lagna and 10th meridian cusps. Once you get these you may evolve all the other things you need from these primary positions. Thus you see that these primary positions are the fundamentals in Horoscopy.

How to fix up the above fundamental elements

You have two ways open to do this –

1.0 Panchanga method
2.0 Ephemeres method.
3.0 Computer Method

We shall first deal with the Panchanga method. Here a warning. You must refer to a Drigganitha Panchanga which tallies exactly with Lahiri's Nirayana Ephemeres or Raphael's Sayana Ephemeres (adopting the Ayanamsha of the Govt. of India like one of Lahiri).

Although the following methods are of no use today , but I suggest my readers to study these also. Today with the help of computer one can know in minutes details of planets and different division charts. I am using Computer method to know position of planets and found Parashara software quite useful. But it is not good for D-11, D-16, D-5. One may have to prepare it. One can know very easily balance period and running period of Disha.

Let us now take up Panchanga Method.

As you are already aware a Panchanga recites the Thithi, Week-day, Start Yoga and Karana that rule the day of birth with Ghatis

34

and Vighatis stated against them. They indicate that those elements rule that day till the end of so many Ghatis after the day's sunrise. In some Panchanga it is also shown in Indian Standard Time. Here please note that the Ghatis etc., shown are from the time of sunrise of the place for which the said Panchanga is prepared and this should not be taken to be the same for all places but if shown in terms of IST it holds good for all places where IST is followed reckoning the birth time also in IST.   If you follow a Panchanga prepared for a place of different Latitude and Longitude from the one you are born then necessary adjustments should be made if to be counted in terms of Ghatis from sunrise. So to avoid all the complications convert the Ghatis into IST and work. To find out the Lagna these elements at the birth time you proceed as follows.

At the above birth time the Thithi is Shasti as Panchami ends at 11 hrs. the birth time being 17.30 hrs. The Star is Bharani as it rules till 19-34 hrs. Yoga is Vyagata as it rules between 6-20 and 27-5 hrs.

To work out their actual positions at birth time you must take the total duration of these elements or a day. How ? For Shasti Thithi you must find out the total span of Shasti. To do this you must count from the ending time of Panchami to the ending point of Shasti.   To find the total span of Bharani Star you must count from the end of Ashwini to end of Bharani.   To know the end of Ashwini star you must look up a day back, Ashwini ends at 21.30 hrs. on 7-9-63.   So count from this point to 19-34 hrs. on 8-9-63 (end/of Bharani star).

To work out the span of Vyagata Yoga count from 6-20 hrs. to 27-5 hrs. on 8-9-63 (being the beginning and end points of Vyaththa Yoga).

Thus you find the total durations o these elements. Then for the birth time calculate by Rule of Three Proportional method the exact point. To know the weekday it is easy as the week commences from sunrise of the day. Thus you have now understood the method of evaluating these elements.

Next how to find out the positions of Planets

Moon's position is to be calculated from the exact position of the birth star as calculated above. To find the positions of the other planets follow the following method.

In the Panchanga the star Padas of all planets will be given whenever there is a change in their motions. Looking rom the birth time of the day and going back find out the immediate proceeding star Padas of the planets and they will represent the positions of the planets. Once you know their Padas it is easy to locate them in the Rasi chart and also in the Navamsha chart as explained in lesson 2. Navamsha charting follows later.

By equating these exact proportions in Padas worked out from finding the total span of each pada in which the planet is and the time elapsed in that also find out the exact degree positions of planets but t this is very cumbersome some cases say of Sani you will have to work out for months to find a span all these Ephemeres is best. This is how to find the planets in terms of our old school of astrologers were satisfied with the mere pada chart. In some Panchangas the Chara Padas of Sun are not given specifically, instead as the kind of Rain starting which is nothing but the starry part The names of rains we speak of is no other than the Chara Padas of Sun.

How to find out Lagna

At the end of the day or on the next or the previous day the Rasi in which the Sun has passed is shown (Rasi Bhukti). Out of the to this Rasi find out the balance and this will be the balance of Lagna

.Thereafter go on adding the spans of the successive Rasis till you get That Rasi will be the Lagna Rasi at birth time. To find out the exact point out of the total span of that Rasi find out how much has elapsed at birth proportional methods treating one Rasi as either 9 parts you can find out of Lagna which is called Navamsha pada or by treating a Rasi as 30^0 you out the position in degrees and minutes. See the example cited on PP 10 Here you have to note an intricate point. In considering the spans of Rasi you must take the spans (the moving spans as per new and modern allotted to particular latitudes of the place. For reasons already explained balance of Lagna at sunrise will slightly differ if you refer to a Panchanga different place but for all practical purposes unless it comes at any two lagans you may consider a Panchanga worked out for the from birth .

Special Lesson No. V-A for 1980

This is the most important lesson dealing with the subject of Astronomical part of Astrology, Though these have been fully dealt with in our text as some felt difficulty to follow them for want of that year's Ephemeris we are now preparing this special lesson for each year as you will be supplied with that year's Ephemeris. The Ephemeris referred to is Lahiri Ephemeris. If in this lesson there be any clerical error please get them set right by the general principles stated by us. We have selected a harmonious case in which the Cuspal and Bhava charts do not differ from Rasi chart. When the sign containing the 10th Cusp does not fall in the 10th sign from Lagna sign differences arise. Even then follow the same principles narrated in this lesson in the matter for the cited example as the general theory is the same. You may in such cases see more than one cusp falling in one sign and some signs going without any cusp. With these general hints we go to the subject proper.

How to use Lahiri Ephemeres and what to note

For purposes of illustration we cite Lahiri Ephemeres for 1980. The pages cited in this year's Ephemeres may slightly differ from other year's ephemeras by one or two pages which should be adjusted by you.

Read the Explanations on PP. 3 to 5. Note the two pages allotted for each month giving the daily positions of planets at 5.30 A.M IST. For example on PP.

38

10-11 that for January 1980 is given. Col, 1 is Date and Weekday, Col. 2 is Sidereal time at 12 Noon LMT for all places to be used for finding Lagna and Tenth cusps., Col. 3 is Longitude of Sun, Col. 5 is Longitude of Moon. On page 11 are the longitudes of Budha, Sukra, Kuja, Guru and Sani in order and on the last col the longitude of Rahu which may not be taken being True Rahu as that does not fit in for astrology and instead take Mean Rahu positions given on PP. 35 once in 10 days. These are the only things to be used. Leave off the rest and this runs till 33 page. On page 34 the periods of Combustion is given. PP. 35 gives periods of Retrogression of planets. Next note PP. 44 to 47 giving the durations of Thithi, Nakshatra and Yoga which is nothing but Panchanga for the year given in Railway time. Sometimes it is beyond 24 hours say 27 hours which means 3 A.M. as he reckons till 5.30 A.M. which will be 29-30 PP. 48 and 49 give the timings of entry of planets in next sings and PP. 50 gives timings of planets entry into next stars. Next see **PP.** 72-73 for table of Dashas. Next see PP. 76-77 for table of sunrise given in in LMT for required latitudes. Lastly use the abridged Table of Ascendants to find Lagna and Tenth Cusp. For notations used in this Ephemeris see his preamble. Leave off other pages as we do not want them. With this advice *we* go to the subject proper.

Illustration cited in this lesson

Birth at 1.30 P.M Indian Standard Time (IST) on 15th January, 1980, at Bangalore (12N58-77E36).

Sunrise

As this remains more or less the same for all

years it is enough if you use any one year's table. See table of sunrise and Sunset on PP. 76-77. Please note that they are in LMT measures depending on the latitudes (Lat) of places. Being an abridged table worked out for only few Lats till 29^0 and for difference of 4 days to find for the required Latitude and Date work by Rule of three measure by taking up the abutting latitudes and Dates. If you want for more than 29° Lat see the rate of difference between 26 and 29 Lats for the given date and work out by proportional method. *In* this example under Col Lat. 13° sunrise is given as 6.29 A.M. on 13.1.80 and as 6.30 A.M. on 17th (both under LMT measures) working on average and proportional basis sunrise on 15th is 6.29 and 1/2 A.M. (roughly 6.29). If you wish to convert this into IST add 20 minutes being the longitude difference for Bangalore as explained below when Sunrise becomes 6.40 A.M. IST.

Local Mean Time VS Indian Standard Time (1ST)

Each country follows a zonal standard time for some local convenience and India follows Standard time which is 5and 1/2 hours in advance of Greenwich Mean time (GMT). The LMT of a place situated east of Greenwich will be in advance of GMT worked at 4 minutes per degree of the longitude of the place (Long) and for western long. at 4 minutes per Ion# less. As Bangalore is 77E36' (east of Greenwich) we get the long difference as 310' or 5 his 10 minutes in advance of GMT to LMT Bangalore. As 1ST is fixed as 5.1/2 hrs. in advance of' GMT to get 1ST from GMT Bangalore add 20'. Conversely to get LMT from (IST) deduct 20" Thus given in one measure we can evaluate the other.

To calculate Lagna and Tenth Cusp

To do this we need the Sidereal Time (Sid. T). Do not ask what it is but simply follow what we say. On PP. 10 under Col. 2 against date 15th the Sid. T is given as 19 hour 36' (leave off seconds in all our workings). Please note that this is for 12 Noon LMT for all places on earth irrespective of Lats, Also note that the motion of this Sid. T is 24 Hrs. 4' a day. For our purposes we take it roughly as one hour for every clock hour. Now find the difference of time lapsed from Noon LMT to Birth LMT If birth is in advance of Noon add and if before subtract from this Noon Sid. T, In this case birth is 1 hr. 10' in advance of LMT Noon and so adding this we get 20 hrs. 40' as the Sid. Time at birth. If you want a more accurate one work out at 24 hrs. 4 minutes a day. As it is after all a negligible quantum that will not affect our readings as we apply birth corrections to the furnished time.

To calculate Lagna

Till you possess the enlarged edition of Table of Ascendants of Lahiri or any other author confine to the Abridged one on PP. 84- 85 where measures are given for limited number of Lats, and for spans of 10° Lagna and Tenth Cusp. Note that the working of Lagna depends upon Lats of places while the 10th cusp stands for all places on earth irrespective of Lats. To work out for the desired Lat of a place work for abutting Lats. and strike out by proportional method for the required Lat. as done for sunrise.

In the present case, on PP. 84 under Col. 13° Bangalore, against Sid T of 20 hrs. 29' Lagna is 20° Mesha

(Vide Col.1). For 21 hrs. 6' it is 30 Mesha. Thus for a difference of 37" Lagna moves by 10^0. So for a difference of 17 Minute Lagna moves by 4' 3' Adding this to that of Sid T 20 Hrs. 29' we get Lagna as 24^0 36' of Mesha at birth time, say roughly $25°$.

To calculate the tenth cusp

This is easy as it is common for all places on earth irrespective of Lat and Long. Look to col under Tenth House for all places on PP. 84. Work out for the Sid. T of birth by proportional method as done in the case of Lagna. For Sid. T of 20 hrs. 23' the 10th cusp is 10^0 Makara and for 21 hrs. 4' it is 20° Makara. Thus for a difference of 41' of Sid. T the tenth cusp moves by 10°. So for a difference of 23' it moves by 5^0 40'. Adding this to that of 20 hrs. 23' we get the tenth cusp as 15^0 40' of Makara, say roughly. 16^0 .

Other Cuspal points

Please note that by cusp is meant the central point of Bhava. As Lagna is at 25 Degree Mesha the 7th cusp falls at 180^0 apart which is 25 Degree Thula. Similarly as the 10th cusp is at 16^0 Makara the 4th cusp will be at 180° apart i.e. a. 16^0 Kataka.

To find the other Cuspal points work as follows: -

Divide the distance from 10th cusp to 1st cusp (Lagna) into 3 equal parts and locate the trisected points between the 10th and 1st Cusps which will be the 11th and 12th cusps. Similarly locate the trisected points from 1st to 4th to get the 2nd and 3rd cusps Thus you get the cusps from 10th to 4th. 180^0 apart from each one of them will be their opposite cusps as :-1 - 7, 2 - 8, 3 - 9, 4 - 10, 5-11 and 6-12 respectively.

42

In this example, the distance from 10th to 1st cusp is 99^0. 1/3 of it is 33^0. Add this to 10th cusp at 16^0 Makara we get 19 Degree Kumbha as the 11th cusp. Further add 33 Degree to this 11th cusp to get the 12th cusp at 22^0 Meena. Further by adding 33^0 to this 12th cusp you must get the 1st cusp at 25 Degree Mesha. Similarly the difference between the 1st and 4th cusps is 81 Degree. 1/3 of it is 27^0. Adding this to 1st cusp we get 22^0 Vrishbha as the 2nd cusp Further adding 27 Degree to this 2nd cusp we get 19^0 Mithuna as 3rd cusp. If you add further $27°$ to this 3rd cusp we must get the 4th cusp at 16 Degree Kataka. Thus we have located Cuspal points from 10th to 4th. 180^0 apart from each one of them will be their opposite cusps.

Bhava Spans

This is needed to fix up Bhava chart. A Bhava ranges from the midpoint between its previous cusp and itself as starting point of that bhava having its center at its Cuspal point and ends at the midpoint between itself and its next cusp. Some mislead saying that a bhava commences at its Cuspal point and ends at the next Cuspal point. This is wrong. In this example the first bhava ranges from 8 and $1/2^0$ Mesha to 8 and $1/2°$ Vrishbha with its central point at $25°$ Mesha. All planets falling within this range are supposed to be in 1st bhava. In this way work out the spans of all bhavas and locate the bhava positions of all planets. Then marking any sign as 1st bhava (preferably Lagna sign for some convenience) place the planets in the sign counted from 1st bhava sign equal to its bhava number. This is Bhava Chart. This chart is used only to know in what bhava a planet is and for nothing else. Some value this chart and speak of aspects, conjunctions and sign positions which is absurd.

To work out Planetary Positions

Ephemeras may give the positions of planets once a week, once in **4 days** or. daily at specific times as 5-30 A.M. or 5-30 P.M. or midnight-all in 1**ST** timings. That does not matter much for us. What we need is to find out the daily motion of a planet which may be calculated considering the abutting dates. Next find out the interval between the given date and time and the birth date and time and work out the total motion of the planet for the said interval and add it to the original position of start if the planet is in direct motion and subtract if retrograde which may be seen looking at its positions on the said two days. Please note that as the positions of planets are given for IST timings the birth time should also be in IST measure to find the said difference.

In the cited Example

Date	Sun	Moon	Budha	Sukra	Kuja	R.Guru	R.Sani
15th	9-0-24	7-22-49	8-26-23	10-05-01	4-21-46	4-16-4	5-3-23
16th	9-1-25	8-6-30	8-28-0	10-6-14	4-21-46	4-16-0	5-3-22

Daily Motions

	61'	13-41'	1-37'	1-13'	Nil	-4'	-1'

Total motions for 8 hours (5 – 30 A.M. to 1 – 30 P.M.)

	20'	4-34'	32'	24'	Nil	-1'	-1'

Actual Positions at Birth Time

	9-0-44	7-27-23	8-26-55	10-5-25	4-21-46	4-16-3	5-3-22

Positions of Rahu and Ketu

Though daily positions of Rahu are given on PP. I1 as it is True Rahu position that does not fit in for predictions please take the Mean Rahu on PP. 35, given for once in 10 days. As the daily motion of Mean Rahu is 3.18' a day note its position given for the date prior to birth date and work out its motion from the given date and time (5-30 A.M) to birth date and time and subtract it (as the motion of Rahu and Ketu is always retrograde) from the previous portion. As Rahu and Ketu is always at 180^0 once you locate Rahu 180^0 apart will be Ketu.

In this example, on 11.1.80 Rahu is at 4-7-47 and on 21.1 80 at 4-7-16. Thus for 10 days he moves by 31' and so for 4 days and 8 hours he moves by 13' 4'-say 13'. Deducting this from the position on 11th we get the position of Rahu at birth time as 4-7-34 and Ketu at 10-7-34.

Balance of Birth Dasha

This depends upon the positon of Moon at birth. Refer to Table on PP. 72-73. As Moon is at 7-27-23 i.e. 27^0 23' of Vrischika falling on Jyeshta 4 whose lord is Budha it is Budha dasha at birth. In the table against 27^0 20' the balance of Budha dasha is Yrs. 3-4-24 (last col). For the balance of 3' see proportional parts at the bottom PP. 73 under col. Budha against 3' which is 23 days.

As the balance of dasha goes on decreasing as the position of Moon increases deducting this we get the balance of dasha of Budha at birth as Yrs. 3-4-1.

May God bless you.

LESSON No. 6

In lesson no. 5 you have understood the method of erecting the Rasi Chart fixing the Planets, Lagna and Tenth Meridian with the aid of Panchanga and Ephemeres. In this lesson you will know the mode of erecting the Cuspal Chart and Bhava Chart which are very necessary when we deal with the ownership and situation of planets. This is described in detail in pages 95 to 101 of Pt. I. Please go through it in detail and supplement it by this lesson.

How to fix up the Cusps of Bhavas

Here there is difference between the Western and Eastern Methods. As per the Western system the cusps of the 10th, 11th, 12th, 1st, 2nd and 3rd all vary and their opposite bhava Cusps are situated at 180° apart. But under the eastern system we first of all fix up the Lagna and 10th cusp and then work out all the other cusps. Under our system we divide the span ranging from the Ist cusp to 4th cusp and likewise that of from 10th cusp to 1st cusp into 3 equal parts and locate the Cuspal points of 2nd, 3rd and likewise of 11th and 12th. Thus we get the 6 Cuspal points from 10th to 4th. The other 6 cusps are got by adding 180° to its opposite. For example add 180' to 3rd cusp you get the 9th and so on. Thus in our system you have to find out the first six cusps. The rest will be located automatically at 180⁰ apart. Thus erect the positions of all the cusps of houses including Lagna and this is called the Cuspal Chart.

Here you may experience some peculiarity. By crude and wrong method of counting the Bhavas from Lagna Rasi you get each one cusp in one Rasi ; but by this correct method you may see 1, 2 or even 3 falling in one Rasi or sometimes no cusp at all in a Rasi. When the 10th cusp falls in a Rasi which is 10th Rasi from Lagna-Rasi then reckoning the cusps of houses from either the Rasi Chart or Cuspal Chart will be the same. It is only when the 10th cusp varies from the above 10th position that there will be dislocation of cusps.

How are these Cuspal Charts useful.

Whenever you speak of the lordship of planets (Adhipathya) these Cuspal positions are of great importance. A Bhava Lord is one who owns the Rasi in which the cusp of the bhava falls. Thus to find the lordship of planets Cuspal Chart is important.

For example, for one born in Dhanur Lagna If the 10th cusp falls in Thula the lord of 10th becomes Sukra instead of the time old conception Budha. For, the 10th cusp is dislocated from the usual Kanya Rasi to Thula Rasi in this case and so the result. So you are now convinced that counting from mere Lagna Rasi is fake and highly erroneous and that is why the readings of some of the astrologers go wrong when they touch up those dislocated bhavas. For heaven's sake please do not follow them. Leave it to the half learned astrologer.

What is Bhava Chart

You speak of a planet as owning a particular bhava being situated in a particular bhava. The old school of astrologers who are not mindful of this subtle difference count all from the Lagna Rasi but that is not correct. You have known about the finding the lordship of planets. Next you must know to find out their positions meaning in what bhava they are and to know it you must erect Bhava Chart.

47

How to erect a Bhava Chart

To erect the Cuspal Chart it is enough if you know the Cuspal degrees of houses but in the case of Bhava Chart you must take both Cuspal degree and the longitudes of planets. First you must fix up the Bhava-Spans.

What is Bhava Span

The cusp is the central point of a Bhava having its entire range spread **over on** either side of the Cuspal point- before and after. To find the range on its fore and prior sides the following procedure should be adopted. To find the fore-span locate the mid-point between the cusps of that bhava and the one before it. The distance from this point to the Cuspal point is the prior part of this bhava. Similarly fix up the midpoint between the cusps of this bhava and that of its next. The distance from the cusp of this bhava to that mid-point shall be the after-span of this bhava. Thus a bhava span commences from the previous mid-point to the after mid-point as calculated above its center being at the exact Cuspal point. Under normal circumstances a bhava spreads over roughly 30 Degree with its central point at the Cuspal points with a spread of about 15 degree on either side of the cusp. Thus you have been able to locate the commencing the central and end points of a bhava. Likewise work out all the bhava spans. For example to work out span of 4th bhava the starting point is got by finding the mid-point of 3rd and 4th cusps, its ending point is got by striking the midpoint of 4th and 5th cusps, the central point being cusp.

How to fix up Bhava Chart

Having thus found out the entire range of a bhava span find out in what bhava span a planet is situated. Then say that planet is in that bhava. Mark the cusp of Lagna in any Rasi and place the planets in the number of houses equal to the situated bhava as found above reckoning from the Lagna Rasi (counting in terms of Rasis). This will be the Bhava Chart. For example for a Dhanush Lagna birth with Budha in 8th bhava span put Lagna in Dhanush (you may put Lagna in any Rasi as it is only the counting from Lagna Rasi that we need here and not the Rasi character) and count the 8th Rasi from the Lagna-Rasi you have chosen i.e., in this **case** Kataka. Beware, do not read here that Budha is in Kataka. You must only say that Budha is in 8th bhava and nothing more. Likewise workout for all the planets and that will be the Bhava Chart. After a little experience you will be able to say in what bhava a planet is situated by merely looking at the Rasi Chart itself the clue being the difference in degrees from the Lagna degree. Till then follow this strict method which is always mathematically correct. In bhava Chart it is merely the number of houses from Lagna Rasi that is to be counted and nothing else as already stated. Aspects or conjunctions should not be reckoned here. Some err on this side but you should not.

How to use the Cuspal and Bhava Charts

You have to use the Cuspal Chart only to find the lordship of planets as stated before and from the bhava chart find out the bhava in which a planet is situated. It has been the time old practice to read both the lordship and situation of planet from the Rasi chart itself. He goes lucky when there is no difference between the Rasi chart and Cuspal and bhava charts, but when they differ the predictions miserably fail. It is by ignoring this fundamental point that many an astrologer falters but you should not after knowing this secret We are in right time to pull you out of such confused minds

Dashas

We have so far dealt with all aspects of horoscopic epigraphy except Dashas which we now touch up here. Even so there are innumerable systems of Dashas narrated by different authors as Varahamihira. Parashara, Jaimini, Tajak, Astakavarga, Kalachakra, Astotthari, Vimshottari etc., and some of the recent inventors go to the extent of suggesting 'Aladdin's Wonderful Lamp' such as so and so Dashas to be followed if such and such planet is more powerful, say most Powerful of all, in a horoscope. Some suggest to consider Lagna star instead of moon star in reckoning Udu Dasha, Some suggest the year as being 360 days and others as 324 days. There are also some suggesting quite a new topsy-turvy order in the sequence of even Udu Dasha. We can only say that all these are due to their failures in the proper handling as per any system. We shall not waver nor falter like them but most candidly and emphatically say that in Kaliyuga Udu Dasha is the only way—the correct and easy way, that will be found to be cent percent applicable in all cases and at all times. So I have firm faith in Udu Dasha and reckon the year as being 365 days (roughly). We say roughly because the correct reckoning is to count the time taken by Sun to complete one full revolution in the map as one year and this more or less is near the English Calendar year.

How are Udu Dashas Reckoned

On page 103 Pt 1(Astrology made easy), the triplets of stars with their dasha lords and dasha years are given. Please study the same and try to remember them. You must first know the star in which Moon is situated at birth. Then the dasha corresponding to this star rules at the birth time till the duration of its balance as birth to be calculated as below and after that period the successive Dashas run for the periods assigned to them the order of Dashas being Sun, Moon, Kuja, Rahu, Guru, Sani, Budha, Ketu, Sukra with heir periods being in years 6, 10, 7, 18, 16, 19, 17, 7, and 20 years respectively.

How to find the balance at birth

By Panchanga or ephemeras find the total duration of the birth star and the period that still remains to be passed in that birth star. By proportional methods find out the number of years corresponding to the above period to pass in the birth star in proportion to the total duration of the birth star reckoning the number of dasha years of that birth dasha. For example for one born in Pubba star when its total span is 26 hours and from birth time the balance of star runs for another 15 hours the birth dasha balance will be 15/26x20 yrs. (Sukra dasha being 20 Yrs. as the lord of Pubba is Sukra). Thus the balance at birth is Yrs. 11-6-14 of Sukra Dasha. If you find out the position of Moon at birth in degrees and Minutes the work is simplified as follows. Suppose Moon is at 12°40' in Kanya at birth then Moon will be in Hasta I padama. Hasta star commences from 10 degree in Kanya and so Moon has passed 2°40' in Hastha star. There is thus a balance of 10°40' to pass in Hasta as the total span of the star is 13°20'. Working by proportional method, for 13°20' the dasha years is 10 (Moon dasha period). So for 10°40' how much will it be--converting all into minutes we have as follows. 640/800 X 10 Yrs. - 8 Yrs.

P.S.—*A* more ready reckoning can be had looking to the Table of Dasha Balance in Lahiri Ephemeras (Annual) on pages 68-69. Against 12°40' under col. 1 under Moon Dasha under col. 3 you get 8 Yrs. at a glance.

Sub-periods or Bhuktis
How to calculate Bhuktis

To find the span of a Bhukti under a dasha multiply the dasha years of the dasha lord by the dasha years of the Bhukti lord and divide it by 120 (total of all Dashas). You thus get the span of the Bhukti periods under that dasha. Example Budha Dasha Guru Bhukti will be got by multiplying 17 (Dasha year of Dasha lord) by 16 (Dasha years of Bhukti lord) and dividing by 120 (total of Dasha-) i.e. 17x16/120=Yrs. 2-3-6. There is a short and quick method of arriving at the same result as follows. Multiply Dasha years of dasha lord by the dasha yrs. of Bhukti lord and from the product so obtained remove the unit digit and reckon the rest as months. Multiply the unit digit by 3 and reckon it as days. The two will give the Bhukti period. Example, in the above case it is 17x16=272. Removing the unit digit 2 we are left with 27 and these are months. Multiplying the unit digit 2 by 3 we get 6 as days and so Budha Dasha Guru Bhukti will be 27 months 6 days or Yrs. 2-3-6 (2 years 3 months 6 days) as obtained above. This short-cut method is based on mathematical trick and nothing else.

In the same way you may calculate the period of Antharas lord treating Bhukti lord as the Dasha lord and Antara lord as Bhukti lord but in this case take the total span of the dasha lord as the span of the Bhukti under which we wish to find out the Anthers. Suppose we want to find out the Moon Anthar in Guru Bhukti of Budha Dasha which will be Yrs. 2-3-6x10/120. To handle the effect of Dasha itself is difficult and more troublesome it is with

Bhukti reading and really a hazardous deal with the Anthar reading unless you are fully conversant with the gift of prediction. Till then we advise you to confine yourself only to Dasha Bhukti readings. To make the work more easily there are again readymade table of Dasha Bhuktis which you may refer for quick handling.

Thus you have learnt in this lesson all the fundamentals of horoscopic Epigraphy. Please digest these and be prepared to follow the next lesson on Shadbala.

Now a day's software's are readily available. These calculations is only one click away. One writes date of birth, place of birth and time of birth and one can get balance dasha, degree of planets, stars and dasha of planets up to Antra Disha and so on.

LESSON No 7

SHADBALA

Chapter IV of our part I deals fully with this aspect of Shadbala. Here you must specially note the difference between Shadbala and Shadvergabala-the former has to be judged in the Rasi chart alone while the latter has 6 be decided by the several Amshas or Vergas what we later on call as Division Charts or Phala Kundali in shorter notation called D-charts. Thus you see that Shadvergabala is the sum total strength arising from the six Vargas or D-charts. Viz, D-1, D-2, D-3, D-9, D-12 and D-30. There are some who reckon Dasha Vargas i.e., 1, 2, 3, 4, 7, 9, 10, 12, 16 and 30 Vargas. Under the theory of Division charts we have propounded that each D-chart speaks of only singular effects and its magnitude has to be judged by the power of planet arising out of `Shadbala' in that D-chart, So is evident that no useful purpose would be served by finding out the sum total strength of Varga bala which will be helpful only to find the comparative strengths of planets and never their nature of effects. So it follows that even after hard labor put forth for evaluating the above said Varga balas we cannot make out its effects on any one particular aspect as that would be the sum total strength of different aspects which has no meaning. So we shall ignore the working of Shadvergabala of a planet and concentrate ourselves solely to Shadbala.

What is Shadbala

It means six fold strength. 'They *are* 1. Kalja, 2, Chests, 3. Ucchaja, 4. Dik, 5. Ayana and lastly 6. Sthana. For details please read PP. III-112 Pt. I. Out of these we shall confine only to four of them in addition to Vargottama bala (Position of a planet in the same Rasi both in Rasi chart and D-chart). These four are 1. Vakra bala 2. Uccha bala, 3. Digbala and lastly 4. Sthanabala. In the book Sthanabala is not explained and that is no more than a planet being in own, friend's, neutrals or enemy's house of which we chiefly prefer to consider only Swatchetra Bala-one in own house leaving off the rest as they do not affect the position too much. Just as you evaluate the magnitude of effects in Uccha position likewise weigh the decreased effects when they are in Neecha and Asta-very near the sun. In the case of Asta planets one exception is found under our New Techniques. If Sun be the birth yogi then a planet in proximity of Sun will instead of causing bad will give yoga to the native. This is what we have practically experienced in all horoscopes.

Be it definitely understood that a planet endowed with Shadbala will only enhance the magnitude of its effects. So you should never attribute good or bad by the mere reason of ˙its strength The nature of the effect whether good or bad has to be judged from other factors which will be explained later.

Caution

Sometimes it may so happen that even though a planet is endowed with some of these Shadbala in a Rasi Chart he may not prove to be good at all. On the contrary he may give adverse results. It is said by Daivaganas that a planet even with such Shadbala will be of no avail if posited in Dustanas. Yet sometimes you may experience wonderful results even under such a condition in a Rasi chart. Similarly a planet weakly posited in Rasi chart may give excellent results. It is such anomalous situation that has baffled many an eminent astrologer who in dismay tries to coin a theory of his own which again fails when applied to another chart when he again gets back to the old theory—thus himself being tossed hither and thither-ending in utter confusion. How to account for all such redundancies ? Yes, we have been successful in winning over such riddles. It is for this reason that we do not attach importance to the Rasi chart alone. We have better modes of judgment as you will know later. In fact even those Daivaganas when they wrote the Dictums did not confine only to Rasi Chart. they had in their mind other factors also. But sorry they have not said in a discussive way though they have said them here and there without a link which we have done to success on an analytical basis. Perhaps they feared that the present set up of degenerated students may not be able to follow those intricate problems of Astrological Astronomy i e , the Mathematical Astrology. Thank God it is made easy-easier than as understood by the Daivaganas. Thus we are in the days of more advantageous situations as most of the intricate mathematical portions are had by Ready-Made Tables. Now all these calculations one can have by computer program or software program.

Let us stop here to take up our next important lesson on 'YOGADHYAYA'.

Om Shanti Om Shanti

LESSON No. 8

YOGAS

The word Yogas used here is different from that used in other contexts. Once when we were discussing Bhava Phal we used this word only meaning a planet becoming Yogic by dint of ownership of Good Bhavas. Again in the context of finding the Birth Yogi caused by the positions of Sun and Moon we have made use of the same word Yoga. Now in this lesson we use the word Yoga only to mean the good effects arising out of the positions of a single planet or number of planets in respect of Sun, Moon, Lagna or the 7th house, Thus you must first understand the different types of Yogas though the same word is used in all the three places.

The Yogas cited in this lesson, though operative throughout one's life, get into actual prominence only during their periods i.e., Dasha Bhukti. When a single planet forms a Yoga as in the case of Ruchaka, Bhadra etc. , then consider its effects as arising during its dasha, When two or more planets cause the yoga then their effects will be felt during the period of any of them and more definitely during the dasha of one and the Bhukti of the other.

We have thus been able to find out the existence of a yoga at birth time and also as to when it would fruitify. You must next be able to gauge the magnitude of the Yoga. To estimate this you must reckon the Shadbala of the planet or planets forming the said Yogas-Shadbala as we have simplified and not in the tough and cumbersome mode of Sripathi. If then as per our estimation of Shadbala a planet forming the yoga be endowed with low medium or high strength then the quantum of the yoga will also be likewise,

All said and done the next question arises. We have seen the existence of some of these Yogas in many horoscopes yet the native will not have enjoyed them even during their Dashas. Why ? The answer is very simple—simple to us but complex to others. You remember our stand that all things cannot be had and judged only from the Rasi Chart (D-1 chart to be known later). So we advise you to apply all these Yogas to the Division Charts (D-Charts to be known later) and then see if even one of them fails in any horoscope—of course during their Dashes.

Though a Yoga exists in a chart if its Dasha does not intervene at proper age you may not enjoy its effect in full.

What is meant by Proper Age

Suppose it is a marriage problem and the planet initiating it has its dasha operating after your 60th age. How does it help one if it is not a case of subsequent marriage as is the case with multiple marriages. Suppose the yoga of education ends before 10th age. Then of what use is his youth for education. Similarly one of professional prosperity operating either before his teens or after his 60th age will be of no use. Thus you see that such Dashas operating at in-opportune will not be conducive to the native to enjoy the inherent Yogas existing in the horoscope.

In Chapter V of our Pt. I a detailed discussion of these Yogas are made which may be supplemented to this lesson. Here we will state only the crux of them enumerating the special points.

Of all the Yogas mentioned in our book you will do well to concentrate your study and attention on a select few going through the rest for academicals interest only. Those of first rate importance are Yogas 1 to 6 and 14 to 17 of the said chapter. Of them the first set of Yogas No. 1-Ruchaka, Bhadra, etc., are formed by a single planet situated in Kendra identical with its own or exaltation house. Here you have to note one exception It really holds good in the case of own house but in the case of exaltation it may not always be true read in the Rasi chart-but it is also true in any other D-chart For as you know already a planet in own house is a protector of the house while a planes in exaltation is only strong to do good or bad depending upon his other characters at ownership etc. Suppose the lord of a Dushtana is exalted in Kendra. Then he does utmost harm to the house he owns and houses he is and aspects. Then what about the sayings of the Daivaganas. They always have said truth, but we do not know how to use their sayings. Apply these Yogas to all the D-charts and see if even one fails. A common opinion is now-a-days current among us that a theory applicable in majority of cases may be embraced as naked truth We differ here. If it should be a science a theory should be applicable in all cases without exception We always state such theories that are applicable in all cases without exception and leave off the disproved or partially proved ones for further research.

Next, Yogas 2 to 4 *are* formed by two planets situated in either or both sides of Sun, Moon, Lagna or seventh house, In fact this may be generalized to all cases of planets being posited at equidistance from Lagna such as 1-7, 2-12, 3-11, 4-10, 5-9 and 6-8. Of course their effects are to *be* read out during their periods, according to nature of those planets —Benefics or Malefics by Nature only.

How to read these effects

If Natural Benefics are situated it is said to, form good, 'Sankhya Yoga' giving auspicious results while in the case of

Natural Malefics only bad has to be read out. Please note here that only the natural qualities of planets to be considered and not their functional trait by ownership of Bhavas. If both the Benefits and Malefics form such a yoga then you have to read mixed effects as warned before. Please always use these Yogas in Division Charts only.

Notes

Texts are silent on the point of Sun and Moon in such positions but on research we have found that they play better part in doing well than the other Benefics. In fact we have experienced that Sankya yoga caused by sun and moon have proved to be the best of the lot. One another point to note here is that in such positions even Malefics in own, house do good and definitely in the case of a planet who is the Birth Yogi.

Yoga No. 5 (Adhi Yoga)

It is formed by three Benefics in 6, 7 and 8 from Moon as all say. But we prefer to count it from Lagna. In fact all the Yogas accounted for by others from sun or moon we would prefer to count always from Lagna.

Yoga No. 6 (Gajakesari Yoga)

This is formed by Guru being in Kendra from Moon. This is perfectly true when seen in our D-charts but you must note a subtler point here when you read from a Rasi chart. For real Gajakesari Yoga Guru should be functional benefit situated in positive house and Bhava. It means that he must be the owner of good bhavas and situated in good bhavas and Rasis. Please note that the counting of Kendra should be done in terms of Rasis in the Rasi chart and not in Bhava Chart. In D-charts it is always the distance in Rasis where there is no ambiguity. The more ordinary natural benefic nature of Guru may give him a place in the society he is but for real prosperity its functional trait should be good. For example a hotel server having this Gajakesari yoga with functional malefic guru will try to boss over his master but he will ever be a server. If on the other hand that Guru is also functional benefic someday he will become a proprietor. If you overlook this subtle criteria then you will err. So please note this.

For other Yogas numbered 14, 15 and 16 please refer to our book where they are explained in detail and are self-explanatory.

Finally we wish to impress on you again on the following important conditions for the fructification of the full effects of these Yogas

I.   Their Dashes should intervene in right ages. Otherwise he will be only dreaming of its happening and will feel just like looking wealth bag in a mirror.

2. All these Yogas are to be applied to the Division Charts and the particular effects pertaining to that D-chart read out as being good or bad as the case may be. For Division Charts readings wait till you receive those lessons.

3. Texts have given different names to different Yogas and different effects narrated. In practice *all* those effects will not be felt. If you want to read those Yogas in the Rasi chart alone then do this way. *See* what are the bhavas these planets own at birth and aspect to add to them the natural Karakatwas of the planets. The sum-total will be effects of the yoga. This would be a better reading than merely repeating the

chattering of the book. If on the other hand you read these Yogas in a D-chart there is no doubt as each chart has its own individual effect. For example in the case of Dhanur Lagna birth if any of these Yogas is formed with Budha then he being the lord of 7 and 10 the effects of these houses will be felt and karaka powers of Budha such as education, intelligence, Maternal Uncle etc. If the yoga exists in D-10 the particular effect of D-10 being Profession the professional effect only will be felt and nothing else.

We stop here and let us request you to master this lesson and be prepared to follow our next lesson on Stree Jataka.

LESSON No. 9

FEMALE HOROSCOPY

As proposed in our previous lesson we now deal with the subject of Female Horoscopy in this lesson. Let it be understood that there is not much difference between the handling of Male and Female horoscopes except in a few matters as will be pointed below. If the lady leads an independent life by herself not depending on any one then all that we speak of in the case of male horoscope are equally applicable to the female chart too. In our Hindu society mostly they will be under the care and protection of others when some of the special effects arising out of the female horoscope will have to be made applicable to her Benefactor (Poshakas). Note the word benefactor here. It means that the effects will be experienced by those under whose care she is at that age—if under parents, the parents will experience it and if under husband the husband does so and if under some outsider he will have them.

Then what are those that will be felt by her and what experienced by Benefactors. All those that speak of her personal traits and body such as her body, complexion, character, trait, formation of organs, health, longevity, happiness, mangalyam (womanhood), her relations, her love matters, her personal acquisitions that stand in her name and on body or control and such other matters that relate solely to her person should be ascribed to her alone and the rest such as profession, finance, status, property etc., that she cannot undertake to possess individually should be ascribed to her Benefactor. As stated in chapter VI of PT. I this sort of sorting out the effects is not special only to Female Horoscope. In fact it applies to all cases of dependents such as in childhood, dotage or being an absolute dependent of someone due to individual incapacity for any cause. Note this sort of readings has to be made out only in the case of total and absolute dependents (and not to temporary dependents) who are for a short time under the care of some or being with someone but depending on someone else. For details read our text.

Now let us go to the particular effects of female Horoscopy. The classics say that the Mangalyam (womanhood) has to be read out from the 8th house, children from the 9th, her chastity from the 4th house, husband's prosperity from the 7th and the rest as in the case of males. But by our research there has been some difference experienced here. That 8th house is house of womanhood we shall second but that her children to be read out from 9th is not cogent nor logical and also has not been found to be true. We would prefer the 5th house to 9th house as in the case of males and reserve the 9th house to read something of the husband's prosperity. Regarding the 4th house in many cases it has failed It is only a house of happiness and no morality is attached to this house. In fact the sex morale has to be judged from the 6th and 7th and Venus. These are fully dealt with in chapter VI which may be gone through in detail. Regarding Kuja Dosha there has been a lot of confusion and fuss about it. We will take up later and discuss it in detail. Meanwhile study what is stated in this chapter as pointed out above. All planets whether Benefics or Malefics cause immorality in man or woman. If those planets causing are Benefics and yoga planets it is an advantageous parley, if by Malefics he or she will feel distress. In this way the results have to be judged and not in the stereotyped way of Benefics doing good and Malefics doing bad. Sometimes it may be difficult to estimate some of the results from horoscopy when we recommend

Palmistry which gives definite reading. If a lady has a mole on the right cheek you may conclude that her chastity is at stake. The more prominent it is the more voluptuous she would be. If on the other hand the same mark be on the left cheek she goes lucky in all respects. Similarly if there are too many crosswise lines on an abnormally developed Mount of Venus then also she becomes voluptuous. This is for those who know something of Palmistry.

Regarding the reading of Mangalyam (womanhood) you must be especially careful. Never be misled that it is always Kuja who causes widowhood, in fact all malefic afflict it. Bear in mind that unless those afflicting planet's dasha is in operation those bad effects should not be read. Texts say that though there be affliction-of 7th and 8th houses without the aspect of benefic the position of benefic in 2 will cause death to herself before her partner. Here please note that by benefic only strong Guru alone should be reckoned. Finally this may be also verified on the palm. If the Line of Union on the mount of Budha bends down towards Heart Line the partner dies first. If straight herself will die. If upwards excellent sign of womanhood and prosperity. In this way whenever doubts arise in horoscopy you may look to palm. Then you become a first rate forecaster.

This is all that should be spoken about Female Horoscopy. Get yourself posted with these intricacies on female horoscopy and in the next lesson await the most important, subtle and the most difficult but made easy by us—the subject of Bhava Phalam.

L E S S O N   N o -   1 0

Bhava Phalam Lesson

In lesson No. 6 we have discussed the types of charts as Rasi, Cuspal and Bhava charts and have and also shown the method of working Bhava Spans. Before you study to this lesson you must first master these differences as they play chief role in the determination of Bhava Phalam.

The judgement of Bbava-Phalams different from the judgement of Graha-Phal i.e., the effects caused by a planet. By a judgment of Bhava Phal we get all the sum-total effects of all the portfolios of that Bhava in one's life while Graha phal indicates its entire effects arising from its natural characteristics and functional characteristics arising out of ownership of bhavas, situation in bhava, and Rasi aspects and conjunctions which we generally make use of while reading its Dasha effects which subject shall be dealt with **in** these future lessons. In this lesson we will confine only to Bhava Phalam. In this connection you have to bear in mind that all the sum total effects indicated by the Bhava Phalam may or may not come to happen in one's life for the simple reason that the dasha of the chief planet or planet imbibing that effect may or may not intervene in one's life or even if its dasha intervenes it may not be in proper age of enjoy ability.

Then, is there no way open to predict any particular event from the Rasi Chart alone. Yes, there is way suggested by our Daivaganas. If you want to weigh the merit of any particular event in life they suggest to consider first the merits of the Bhava connoting the aspect and then couple it up with the merits of the planet who is the karaka of that particular effect. Weighing thus both the Bhava and the Karaka you have to come to a definite conclusion of the particular aspect. We have by this. time understood the way of judging the merits of a Bhava. How to estimate the merit of a planet. This is very simple. Consider its position in Bhava Chart, its situation in Rasi Chart, and its strength from Shadbala narrated in Ch. IV. Pt. I. Here you have to note one thing. They say that the condition of a Bhava is of first importance and so if a Bhava fails or becomes weakened (by its lord becoming weak) then all the port folios of the Bhava will fail. In reality it is not the case. We have seen in many cases where even when the Bhava failed some of its portfolios have come up well and it is only some others that have become bad. So how to discriminate this differential effects. If both the Bhava and the karaka go good then that aspect is definitely good and if both go bad then that goes bad. But what happened when one is good and the other bad. This is rather a ticklish problem. If the Bhava is good and the Karaka is bad then bad only prevails with respect to the Karakatwas of it. But if the Bhava is bad and the Karaka good then the event turns out to be good. But the above good or the bad arising out of contrary effects of the bhava and the karaka will be middling and sometimes of mixed character and wholly good or bad. A planet that is badly situated in a horoscope may be bad with respect to some of his Karakatwas but good with respect to others and these depend upon the concerned Bhavas. For example let us suppose that Guru is badly situated in a horoscope. Then the presumption is that all the Karakatwas of Guru are entirely lost. Guru is both Dhana karaka and Putra Karaka. While estimating the bhava suppose the second Bhava (Dhana Bhava) is good while the fifth bhava

(Putra Bhava) is bad. Then the bad of Guru reflects on the point of children wholly and partially on finance.

All said and done, in our experience we have not been cent percent successful in handling this aspect from the Rasi Chart, but we cannot ignore the Dictums of our Daivaganas. If you wish to adhere only to Rasi Chart predictions there is no other go than to resort to this method only whatever be the percentage of success. But if you want to attain cent percent success then you must go to our Division Charts which will be described later.

Next we deal with some more aspects of these Bhava Phalam and try to solve some hazy notions. Here we point out the mere singular points leaving off the rest to be studied from our text.

1. **Magnitude of Bhava Phal**

While judging a planet's effect on a bhava either by aspect or conjunction you must note its distance from the Cuspal point of that bhava—the nearer it is the more the effect felt and vice versa. At the commencement or end of a bhava span its effect is zero so far as that bhava is concerned. Here you must note one thing. This is clear so far as conjunction is concerned. The same rule also applies in the case of aspects. In the ease of aspects the degree of aspect of a planet on a bhava depends upon the distance between the cusp of the bhava and the planet. Of course the first condition of Rasi aspect must be fulfilled. Then only we must look to the degree aspect. The first establishes the existence of aspect while the second measures the strength of aspect. This distinction should be specially marked as otherwise you will be misled by other timely interpreters. Some have commenced to prefer the cusp of a bhava as its starting point which is absurd from common-sense that a bhava's effect is at maximum at its center and Cuspal point. This itself establishes that cusp is the center. In measuring the above stated aspecting strength we have to reckon the degrees progressed by the cusp and the planet in its respective Rasis. If both are situated at the same degree then there is full aspect. Otherwise it varies according to their differences. Suppose the fifth cusp falls at 18° Vrishbha' and Guru is at 8° Makara. Then the aspect of Guru on 5th bhava will be roughly $(18/8)/15=10/15$. Note we have to divide by 15 as it is half a Rasi span within which the strength varies from 0 to 100%. In this connection you should also reckon the Applying Aspect and separating Aspect on a Bhava. When a planet is less than the Cuspal degree in its Rasi then it will be applying and when in excess separating. Applying aspects are more powerful than separating ones. The one indicates progress as age advances and in the latter case it goes on warning as age advances, In the above cited case as Guru is at less degree position than that of the cusp he is Applying to the said bhava and so waxing and more powerful.

2.   Bhava Sandhi

Here there is misconception. It is generally said that a planet in bhava sandhi becomes ineffective. It is not wholly true as/it is commonly interpreted. It becomes ineffective so far as the bhavas that are in-between the planet. But it will positively give the effect of the bhava it owns and the Karakatwas it denotes. For example Kuja for Simha Lagna is in bhava sandhi between 8th and 9th bhavas then you may say that Kuja does not affect in any way either the effects of the 8th or 9th bhava. But he will have to certainty give the effect of 4th and 9th bhavas it owns and of its Karakatwas as younger Coborns, lands, prowess etc. While

70

reading the effects of a planet in bhava sandhi of its owned bhavas it behaves first as if placed in the fore bhava which latterly will have the effect of being situated in the after bhava. In this case Kuja as lord of 4 and 9 causes first bad effect of 8th position and then gives good of the 9th.

Likewise the Karakatwas of Kuja should be read out. Thus you will see that in all cases of Bhava Sandhi the final result shall be in accordance as the next bhava to which it moves will be. Here is a point to be noted. This holds good in the case of planets moving in regular direction. If a planet be retrograde then its ownership and Karakatwas effects will be as per the fore bhava only throughout as by retrogression the planet will get back to the first bhava alone.

3. Position of the lord of a Bhava from that Bhava Rasi

Here again care must be taken in finding the lord of a bhava and his count from that bhava Rasi. As explained before find the Rasi in which the cusp of the bhava falls. That becomes the bhava-Rasi. Then see in how many Rasis the lord of this Bhava-Rasi is placed counting from that bbava-rasi. We have dealt with the effects of a bhava lord being posited in a bhava counted from Lagna bhava (in Bhava Chart only) Daivaganas want us to couple this with his position counted from the concerned Bhava Rasi too. This further confuses one as one way it may be good and the other way it is bad. Then what is to be concluded. Our Daivaganas went on quoting all truths no doubt but they have done in a hap-hazard way without notes of discrimination. If they had done so we would have all become wiser and greatly benefitted. All sayings of all Daivaganas (Not the present day authors) are true but we have to make a same, logical and cogent interpretation of these laws which many a time appears to be

71

contrary and misleading too but really not so. Then how to reconcile this difference. The riddle has to be solved this way. All effects pertaining to the native of the horoscope has to be chiefly read out from the position in Bhavas counted from Lagna and the effect of the relatives connoted by the bhava should be read out from the positions of its lord from that bhava Rasi.

For example, the dasha of a planet endowed with the power of causing marriage occurring at 8th age, one that causing professional good in dotage, etc., does-not help in the actual realization of them. This will be explained at length in our lesson on Dasha Bhukti. Till then we shall confine ourselves to gross Bhava Phalam,

How to judge a Bhava

These are fully discussed in Ch. VII Pt. 1. So you will first go through them at first and then turn to the special hints pointed out in this lesson.

I. (A). First note the planet conjoining with the bhava in question and then look to the planet aspecting this bhava. Here you have to note both the natural and functional traits of planets-the former arising out of their natural Karakatwas (Natural benefic or malefic quality) and the latter by ownership of bhavas. In this connection you should be extra careful in fixing up the Bhava and the aspect. We have already said that the aspects and conjunctions should be read out only in the Rasi Chart counting in terms of Rasis only, and not by degrees as is observed in Western methods. Even in the reckoning of aspects not all aspects of the western should be considered. It is only the fixed special aspects as stated by Parashara that have to be reckoned. According to this all planets aspect the 7th Rasi and planets therein. In addition to this there are some special aspects to some planets. Kuja aspects the 4th and 8th, Guru aspects the 5th and 9th, Sani aspects the 3rd and 10th. Next comes the most intricate problem as to where to look in to these matters. Merely because a planet is in 4th Rasi counted from the Lagna Rasi you should not conclude that this planet conjoins 4th bhava. This is true only when the 4th cusp falls in this Rasi and not otherwise. If perchance the cusp of the 4th bhava falls in the next Rasi i.e., the 5th counted from the Lagna Rasi then this planet will have no conjunctive effects with the 4th Bhava. But for purposes of finding out the situation of this planet it may be in 4th bhava (Bhava Chart). Yet, when we speak of the conjunctive effect on the 4th bhava we have to say that the planet does not conjoin 4th bhava. That it has gone to the 4th bhava will be made us of only to study the effects of the bhavas it owns and no further. This subtle difference in a planet going to a bhava and yet not having conjunctive effect on it looks *a* riddle but that is truth. For example in the case of Dhanur Lagna when the Lagna point is 25° and the 4th cusp at 4 degree Mesha instead of in Meena (4th Rasi from Lagna Rasi) some upsetting positions may be seen as follows. Suppose Guru is at 20° Meena. Though looking at the Rasi chart it may appear that Guru to be in 4th in this case actually he does not conjoin 4th bhava while he conjoins with the 3rd bhava as the cusp of the 3rd bhava falls in Meena where Guru is. The cusp of 4th bhava falls in Mesha and hence no guru there and so no conjunction with 4th. Even in the case of reckoning the aspects of Guru the same analogy should be extended. As this Guru in the Rasi chart aspects Kataka, Kanya and Vrischika those cusps situated in these Rasi and planets there will *be* aspected *by* this Guru. If you work out the Bhava Chart you will see that this Guru is situated in 4th bhava only. As lord of

Lagna in 4th bhava is good to Lagna bhava while as lord of 3 (as 3rd cusp falls in Meena) in 4th Bhava does bad to some of the evil effects of the 3rd bhava. The younger co-burns may become powerful. Similarly the Karakatwas of Guru-issues, money etc., get better being in 4th bhava. This most important point of discriminating between a planet in a bhava and a planet aspecting and conjoining a bhava should be very clearly understood.

How to read the Differential Effects arising from the Natural and Functional traits of a planet

Let us first deal with the natural effects. If Natural Benefics occupy or aspect any bhava (good or bad) they always promote the bhava—give good effects of the bhava in question. On the contrary, if they be natural Malefics they destroy the bhava it occupies or aspects as per our above definition. Here one thing to be specially noted—the Bhava-Karakas should not be in the concerned bhavas, as by the theory of "Karako Bhava Nashaya" some bad will be felt in respect of that particular aspect. For fuller details on this subject await our further lessons. Who is natural good or bad is again a ticklish problem. Generally Sun, Kuja, Sani, Rahu and Ketu are taken to be natural Malefics while the rest as benefice. This is a very broad classification. In certain cases some of these Malefics become Benefics to some aspects. For example Sun in 7th does not harm the partner. Ketu in 12 though may cause all other bad is good for causing Moksha or salvation. Sani in 9 though bad to father and wealth does work well on the philosophical plane of life. Rahu in 5 will cause multiple issues though causing bad to some, In this way you have to note certain exceptions and so you cannot make a sweeping remarks about this. In fact you will see later that Sankhya Yoga caused by Sun and Moon on either side of Lagna is the best of all. After all the natural qualities of planets will only shape the superficial qualities of the bhava while the functional qualities cause the inherent effects of the Bhava. In other words the natural ones blend the character, conduct and superficial traits in man while the functional ones cause the inherent Yogas on the material plane. The one is a - pointer of character while the other prospects in life. As most of us are interested in knowing the material prospects in life than the spiritual side we may even ignore the natural qualities of planets and lay emphasis on the functional traits (Ownership).

Here is a perplexing problem which has not been unanimously and definitely solved by others which we have attempted after long research. Some are of the opinion that the Malefics in bad houses do well to those houses. This is against logic and commonsense. Being Malefics they should harm the bhavas as per general theory. It really signifies that the bad effects arising out of these bad bhavas become still worse by further conjunction or aspect of natural Malefics and also their natural Karakatwas also go bad. For example, Kuja in 6 not only increases enemies and disease and debt but in addition will cause bad to its Karakatwas as younger Coborns, lands prowess etc. For the same reason Benefics in bad houses lessen the bad effects of these bad bhavas but the Karakatwas of these benefice go bad.

For example, Guru in 6th bhava lessens the evil effects of the 6th bhava but his natural Karakatwas suffer such as the welfare of elder Coborns, money, children etc. This is the correct mode of judging the effects of planets Benefics or Malefics, by nature. What is said of conjunction with a bhava equally applies to aspects on bhavas too. Let us remind you again that conjunction or aspect should always be judged in the Rasi Chart or in the Division Charts (which will be dealt with later) of course counting only in terms of Rasis. After establishing the existence of this conjunction or aspect in the Rasi chart then you may consider the strength of aspect by viewing the difference in the degree aspect as western method denotes. So are effects of the Natural Benefics and Malefics. Next we shall deal with the Functional qualities of planets.

I. (B) How to handle the effects of functional Benefics or Malefics.

When we talk of the functional traits of planets we have to particularly note that the Ownership has to be judged from the Cuspal chart and not by the time-old rule of reckoning from Lagna Rasi in order. Similarly the situation of a planet in a bhava has to be judged from the Bhava Chart. Its condition, aspects and conjunctions are to be looked into in the Rasi Chart. Thus there are three things to be noted—the Rasi Chart, the Cuspal Chart and the Bhava Chart. These elements are fully discussed in PP. 93 to 101 Pt, I. By cusp of a bhava is meant to indicate its central point where its effects are at zenith. The lord of a particular bhava is the lord of the Rasi containing the cusp of that bhava. This must be clearly understood, else you may also err like the present day astrologers. Sometimes more than one cusp may fall in one Rasi when one planet may own two consecutive bhavas and thus become the lord of more than 2 bhavas usually found in a normal chart. Sometimes a planet may go without the ownership of any bhava. For example Lagna falling at 29° Dhanush when 10th cusp falls at 0° Thule and 11th cusp at 30° Thule. In such a situation the lord of 10 becomes Sukra and not Budha as per old conception. The lord of 11 is also Sukra as the 11th cusp also falls in Thula. Thus Sukra becomes the lord of 10 and 11 and by working out in full (or by symmetry at glance) you will also find that the lord of 4 and 5 will be Kuja. Kenya and Meena go without any cusp. It is by this method that you will be able to correctly assess the lordship of houses and not by the crude method of taking the Rasi Chart alone as the basis.

What is Situation.

If we merely say that a planet to in such and such a bhava merely counting from the Lagna Rasi it may not in all cases be correct. For true estimation of it you must locate Bhava-Spans. By bhava span is meant nothing but the interval .from the starting of the bhava to the ending point. By situation is meant in what bhava span a planet is situated. We erect a Bhava Chart on this basis and then say that a planet is situated in such and such a bhava. A planet that may appear in the Rasi Chart as being in the 5th house may actually go to the 6th bhava and if he happens to own say 10th bhava you conclude that the lord of 10 in 5 does good to profession while actually he will be suffering in professional matters which is correctly evidenced by the lord of 10 being in 6th bhava and not in 5th. Thus to estimate the correct worth of functional traits of planets the ownership has to be found out from the Cuspal chart while the situation from the bhava chart.

How to read the functional effects of planets.

1. The lord of a good bhava (hereafter called Functional Benefic), if situated in good bhava (seen in the bhava chart) then the effect of the bhava in question will become good but if it is situated in bad bhavas the effect of the bhava it owns becomes bad.

2. The lord of bad bhava (Functional Malefic) gives

77

badly to the bhava it owns if situated in good bhava. But when he is situated in another bad bhava it causes surprise good called 'Viparitha Raja yoga'.

In other words it may be more easily understood this way. Functional Benefics becoming strong being to good bhavas increase the good effects of the good bhavas it owns while in bad bhavas by reason of their weakness spoil the good effects of the good bhavas it owns. Similarly, Functional Malefics in good bhavas, by reason of getting strength increase the bad effete of the bhavas it owns while in bad positions by reason of their weakness cannot do that bad when good only prevails, Let us illustrate this by an example Suppose we have to judge the effects of 5th bhava The lord of this house being functional benefic if situated in good bhava (in bhava chart) will give good effects to the portfolios of the 5th bhavas such as happiness to children, promotion in life, lucky chances arising etc., but if this lord is housed in 8th bhava he may cause death to children and bad in other aspects pertaining to this 5th bhava. Similarly let us consider the case of a functional malefic say 6th house lord. If its lord is in good bhava in bhava chart then by reason of getting more power does more of harm in matters of 6th bhava—more debts, disease and enemies If on the contrary he be in Dushtana he will by reason of becoming weak lessens the bad effects of this 6th bhava. In this way you should judge the bhava effects. So far we have considered only the good or bad effects arising out of a bhava. Further we will take up the estimation of the magnitude of effects—of the good or bad already decided as per above cannons. This has to be judged from the ' Condition.'

What is meant by condition.

By condition is meant its situation in own, friend's,

enemies, exaltation, neutral, debilitation, retrograde or Asta-all to be looked into in the Rasi chart alone for the present which will be later on extended on to Division Charts when you happen to study it. Even here different types of results arise from the different conditions in which the bhava lord is.

1. Functional benefic or malefic in his own house does always goad to the bhava he owns in a greater degree of effect.

2. Functional Benefics in friends, neutral's houses do good but in a lesser degree.

3. Functional Benefics in enemy's or Neecha house or in zero Rasi gives very little well.

4. Functional Benefics in exaltation give well of a far higher order and likewise if he be retrograde.

5. Functional malefic in neutral or friendly house gives bad effects.

6. Functional Malefics in enemy, Neecha, or zero Rasi or Asta becomes weak to give out their functional bad effects and thus good only prevails.

7. Functional Malefics in exaltation or retrogression cause more bad by reason of their attaining higher strength.

In this way all the bhavas have to be considered and conclusions arrived at. Whenever a bhava effect has to be evaluated with respect to a particular portfolio of the bhava the karaka of the said portfolio has to be also considered and then predicted considering both the bhava and the karaka. If you wish to know the financial position of a native you should take both the 2nd bhava and Guru and concluded. The bhava may be found to prosper as per the above rules. If the karaka is weak then that particular aspect of the bhava may suffer while some other aspect may flourish. If on the other hand if the bhava is totally lost then even if the karaka is good he cannot show good regarding that particular Karakatwas. Thus it is established that the condition of the bhava is primary and that of the karaka secondary. This should be specially noticed.

In addition to the above classical rules of estimating the value of a bhava we add some more very important ones which if neglected the entire readings are liable to go wrong sometimes especially whenever these elements, clash with the general cannons. They are the effects of Thithi and yoga. Consequent on a birth in a particular Thithi some of the Rasis go bad and that is termed as Zero Rashi. Similarly birth in a particular Yoga causes some stars and planets acquiring special additional traits. They cause sometimes lifelong effect irrespective of the Dashes that may run. Let us first deal with the effect of Thithi. The Rasi or Bhava that becomes a zero Rasi is just like a Dushtana and its owner has to be ranked as functional malefic. As per general rules the 4th bhava is good and its lord will be known as functional benefic. If that 4th Bhava Rasi (please note the special wording of bhava Rasi. It means that Rasi which contains the cusp of that particular bhava and not the one roughly counted from the Lagna Rasi) becomes a zero Rasi then this 4th bhava becomes Dushtana-like and its lord unless badly situated does harm to this 4th bhava. These will be dealt with in detail in our future lessons. If on the other hand it be already a Dushtana & zero Rasi its lord will all the more be worse and unless he be very badly situated and completely spoiled very bad and even worse results like death, accidents, losses degradations etc., will happen. This is how you should make use of the Thithi effect.

Yoga effect.

Any planet that becomes the birth yogi or the duplicate yogi and more prominently the former will always do good wherever he be placed in the horoscope and whatever bhava he owns. This is one specialty with birth yogi which is not known to any one so far. Likewise, the birth Avayogi with all and any good positions will only cause bad as inimical influences, oppositions and failures in life etc. So the best situation for a birth Avayogi to cause good is when he is spoiled and weakened. He is to be treated just as a functional malefic. Planets in conjunction with these yoga and Avayoga planets or aspected by them imbibe similar characters as in the case of conjunctive and aspective effects of planets. One exception is when Lagna being zero Rasi its lord does not become a functional malefic. Similarly the birth yogi owning zero Rasi still retains its yoga character.

(C) Next see the star in which the lord of the bhava is situated.

1. A functional benefic in the star of another functional benefic does good for the bhava it owns. If he be situated in the star of functional malefic both good and bad will result—well due to ownership and bad due to stellar ship. Of course this should be coupled with the situation of the lord of that Bhava. Please note that this altered effect due to stellar position will be specially felt

during the particular quarter of the Dasha corresponding to the star pads of the planet. For example if Guru is in Aridra 3rd pada and, if Guru be the functional benefic in the horoscope and also situated in positive position then during Guru Dasha good only prevails except during the 3rd quarter of the Dasha i. e., between 8 and 12 years of Guru Dasha when only the effect of Rahu will be felt as the stellar lord is Rahu (owner of Aridra). I am of the opinion that one should study star positions and its effect as per the above theory.

2. Similarly a functional malefic in the star of functional benefic gives also mixed effects as above—bad all through but good during the particular quarter of dasha. If on the other hand if this functional malefic be in the star of another functional malefic very bad results will arise unless as per above cited rules he is very much weakened.

3. Whether functional benefit or malefic if he becomes eclipsed by proximity to Sun (unless that sun is birth yogi) then the functional malefic gives good results while the functional benefic gives bad, with respect to the bhava it owns. If this Sun be birth Yogi then the planet that gets eclipsed imbibes the yoga character and gives yoga. This is one exception we have found out as candid truth which is not known to anyone so far. So please note this.

(D) Next see if any of the Yogas cited in the text (Ch. V. Pt. I) exists with respect to the bhava in question. For example a functional benefic may be in 8th and you may read the result as bad. But if in 6th there be a natural benefic forming shubba Sankhya yoga then the bad will be averted.

In this way you should judge the effects of bhava considering all the above factors very carefully. Though they may seem to be too many after a little practice you will be able to conclude the merit of a planet in a second after glance.

Hence digest this most important lesson before you receive a more striking lesson.

Lesson No. 11

(Bhava Phalam Continued)

In lesson No. 10 we have discussed

1.   The rules to estimate the worth of a Bhava (judged from Rasi Chart),

2.   The ownership of it (Judged from Cuspal Chart).

3.   Situation of that owner (judged from Bhava Chart),

4.   Condition of that lord of that Bhava (judged from Rasi Chart).

5.   Stellar lord of the said lord of the Bhava (judged from the star it is situated in), and finally

6.   From a consideration of the Sookshma lord of the Bhava (which is nothing but the lord of the star in which cusp of that Bhava is situated which is explained later) just akin to the Sthoola lord (the lord of the bhava Rasi judged from the Rasi in which the cusp of the bhava falls).

It is only on consideration of all these factors that you can come to a particular judgment as to the real worth of Bhava and not by the superficial judgment made from the mere Rasi Chart. But in the course of such careful and minute survey you may come across with many conflicting views.

84

Then how to judge definitely. No doubt these conflicting effects found with respect to a Bhava will be true but they will have to be read out only during their Dasha Periods. Even so, it may so happen that some of them not at all come to pass in one's life for the reason of its Dasha not intervening in appropriate time. What is meant by appropriate time is explained in our Lesson No. 10. Thus a gross survey of a Bhava will only give you rough estimate of the results and see that very rarely definite results since a Bhava connotes plenty of things and so it would be difficult to assess the real merit of any particular portfolio of the Bhava.

In the case of bhavas which also speaks of the personal effects of the native it has to be read out this-way first the effects as it is removed from the bhava Rasi will happen and latterly and finally that of bhava from Lagna. Thus it is seen for a native it is the bhava position counted from Lagna that stands in the end. For example for Thula Lagna Moon is in Kumbha. Being lord of 10 in 8th from 10th will first cause change of avocation or disturbance to the one existing. As to whether another will be immediately had we should see its bhavaic position. Moon lord of 10 being in 5th the house of promotion he will jump on to a better position in life. That is what actually happened in the case of an individual during his Moon Dasha. Citing another example if the lord of the 9th in a horoscope be in 4th then being in 8th from 9th it kills father who by reason of being in 4th from Lagna will confer wealth and prosperity to the native as he is a benefic to the native but malefic to father. That because he is in 8th from his 9th his wealth may be first disturbed which latterly comes back to him. This is how you have to reconcile the seeming contradictory sayings of Daivaganas who have remained silent without going into elaboration. Perhaps they must have taught these to their pupils direct which some of their successors have kept them secret.

4. Rest of the matters such as Karaka in Bhava, Bhavath Bhava etc. may be studied from the books.

5. Finally *we* touch up the Sthoola Bhava and Sookshma Bhava.

All these years you know only one lord of a bhava. You will know from us that there is another lord also of a bhava, who plays more prominent part especially when Udu Dasha system is followed. You may call him as Sookshma lord or Duplicate Lord. The lord of the sign in which the cusp of a Bhava falls is called the Sthoola Lord while the lord of the star in which the cusp of the Bhava falls is known as Sookshma lord.

How to use these two Lordships

Sometimes you find that during the period of a planet effects not related to it either by ownership, situation, and aspects or by natural Karakatwas will happen for which you cannot assign cogent reasoning. Then look to the Sookshma lord of that Bhava which will reveal its jurisdiction over the issue.. Thus you will have to prepare two lists-one containing the Sthoola and Sookshma lords of all the 12 Bhavas and the other classifying from the above list all the Bhavas owned by a planet in either capacity. This will help you at glance to read all the effects of a dasha. In chart No. 1 of our illustrated charts published in our text you will see that, Rahu situated in 8th Bhava has caused marriage though he has no direct connection with the 7th bhava. Look to the Sookshma lord it is Rahu. the 7th cusp being situated at 10°57` of Mithuna.

For further details please go through our Pt. I Ch. VII Sec. E. Await our next illuminating lesson No. 12.

LESSON No. 12

In the first eleven lessons we have almost touched up the entire range of elements required for the prediction of a Horoscope. This covers the entire portion of our Text Book N. T. P. Part I. As these elements are of paramount importance in the handling of a Horoscope you must have a clear conception of all these elements. So, before going to the actual predictive aspect we long to have this refresher lesson-covering in a summary way all the subjects dealt with so far. Please note that all chapters that are referred to in this lesson relate to our Part I Text. In the course of our Revision you will also find many a latest research results discovered of late. Thus we are at every moment at research and new ideas and new theories are unfathomed. In this Refresher Lesson only salient points are touched upon leaving the rest for you to study from our Text. Now let us go to the subject proper.

CHAPTER I

Hora

This is very important. It means one hour. It is Hora that forms the Weekday and it is Hora that governs the hour of the happening of an event. Important events generally happen in the Hora of the Dasha or the Bhukti lord, In the case of Rahu it happens at Rahu Kalam only and in the case of Ketu we are still under research. Some say that it would happen during Kuja Hora and others say during the Hora of the planet connected with Ketu or its Rasi lord. Still we have not come to definite conclusions in the case of Ketu.

Thithi

This is very important as it molds the character of an effect. For persons born in particular Thithis some Rasis become Zero Rasis and these will modify the effects of those Bhavas and their lords or those who are in the star of those lords. Those born on Full Moon or New Moon day are not affected by Zero Rasis and unlucky are those born on Chaturdashi as four Rasis become Zero Rasis unless there be counteracting measures in the horoscope to ward off this negative effect by the theory of Double Negative making one affirmative under our New Techniques.

The Thithis are also used by us in the determination of Progeny-vide book on Astrology / Astrology made easy by the Author.

Yoga

This is of top ranking importance in shaping the real destiny of a person. Any prediction made without considering the effects of Thithi and Yoga is sure to fail and will thus bring discredit first to the astrologer and then to the very science. So beware of this hint. The Yoga effects narrated in detail in PP. 13-15 may not be always to be taken to be wholly correct. They are merely general tendencies and not certainties. So you may take them for what they. are worth.

Karana

This is used by only when we consider the tests of progeny till such time as we extend it on to further aspects as and when our researches on the topic reveals with certainty. Thus ends the first chapter of our Text.

CHAPTER II

Graha Samyam

This comes out true in many cases. While reading its effects what are actually stated in the book may not happen. Intelligence lies in the interpretation of those results suited to the native's circumstances. For example if a planet is in Kirita Dharanam Samyam for one in the run for Kingship he will be actually crowned but in the case of others he may catch a new post. Unfortunately even here there are different methods of evaluation. Let there be many methods and alternatives. We have tested them all and winnowed one by one and finally have stood by the one theory we have narrated which has stood the test in many cases. Please note that these Samyam are to be read out for the Dasha Lords during their periods only and not at other times. Remember you have to first judge the merits of the planet from the body of the horoscope; and then this Samyam applied which is after all secondary in nature. Suppose a Dasha Lord is bad in the chart but with good Samyam. The good effects of the Samyam may not come to pass in full degree but their glimpses only may be seen. Also these sayings being of classical days they must be adjusted to modern conditions existing.

Rectification of Birth Dasha

In our experience it is in very few cases say less than 5 per cent of the cases that this rectification is needed. So you may carry on with the balance as found with the Drigganitha Moon as may be found with the Ephemeris or Panchanga we have recommended in our first lesson. It is only when you come across with discrepancies that this correction may be applied. We are at research as to when exactly this should be applied. One thing you should remember in this connection is that any changed effect either due to change in the Dasha Lord or Bhukti lord does never happen on the exact day of such a change as Gochara is the guiding factor to decide the point of fruition.

Family

The definition of family has to be carefully noted as otherwise you may falter in your prediction when you handle the second Bhava that represents Family. To be more definite of the particular family member that may be effected by the Bhava you must look to the natural karaka planet. Suppose the lord of the 2nd Bhava is lost or weakened then some of the family aspects are gone. If suppose Sukra is strong, Guru is weak and Moon is middling then you must say that this wife goes lucky and good, his elder brother bad. and mother of mixed effects. This way the effects of all Bhavas should be read.

Bhava Karakatwas

Things not said elsewhere are said here and grouped in an intelligent and methodical way for ready reference. They are all important. Of them you may remember only those of daily needs leaving the rest to be referred to when needed.

Eighth Bhava

This is a peculiar point in a horoscope. It is both a house of longevity and death and accidents. How to differentiate the two. A planet in 8th except Sani kills the native while the lord of 8 getting weakened also diminishes the longevity while getting powerful and being in good Bhavas increases longevity.

Travels

Note the differential effects in the travels caused by different houses. To predict the land sea or air routes consider the Karakatwas of the planet causing it at the time. In doing so merely the nature of that planet will not do the planet. The Rasi it is, the star it is will have to be considered and the strongest of them to be said. One of the Akash or Vayu Tatwa causes air Travel, one of Tejo Tatwa by all locomotives on land, sea, or air and one of Jala *Tatwa by* sea and one of Prithwi Tatwa on land. Whether the travel is for good or bad has to be noted from the net effects of the planet as causing. good or bad.

Numerology of Planets

Due to the press errors the numerals of all planets are not given in the text. So we give them here The numerals of Sun on to Sani in order in unit digits are— 1&4, 2&7, 9, 5, 3, 6 and 8 respectively. As per Nadi Principles there are specific numbers ascribed to .Stars by which important events are read out in life. We are at it and as soon as we catch it with success we publish them too. For Sun and Moon there are two numerals. When to use what particular number is under investigation. Till then use both. This has to be read with respect to the Dasha or Bhukti lord that rules at the time. All numbers of more *than* one digit will have to be brought down to unit digit by adding all the digits till you get the unit digit. For .ex. $1234 = 1 +2+ 3 +4 =10=1 +0=1$. This way all numbers should be brought down to unit digit and then read out. If one is running Sani Dasha at the ages of 8, 17, 26, 35, 44, 53, 62, 71, 80 and so on he will have a definite effect. What it is should be judged from Sani in the horoscope. It is not at all time these ages of Sani's effects would be felt. It is felt only during these ages when exactly the Dasha or Bhukti of Sani rules. It is only at those ages his effect good or bad will happen. You will know more on this subject on this subject in our ensuing publication on Numerology.

Rahu and Ketu and their Houses etc.

Varied versions are there regarding this aspect. But in the rare Nadi texts we have culled out the Cal fact which is proving true in our research. So what we have said of them about the ownership, exaltation & depression signs may be followed.

CHAPTER III

Ayanamsha

Turn a deaf year to all jugglers on this subjects who slide away from the one fixed by the Govt. of India on the recommendation of the Calendar Reform Committee according to which the Ayanamsha on 1-1-1965 is 23° 21' 48" which is adopted by Lahiri. Simply follow this without entering into further discussion. You will then shine. For it is a bit of truth we are giving you after verification with many horoscopes and the Prasna we have handled when a minutes difference in the position of planets or Lagna will alter the positions. If you still doubt this we will also doubt your faithfulness in your studentship under us and we are sure you will never succeed in correct predictions.

Tatwa Theory

This is very helpful for rectification of Birth time, at the first stage-of course the next and the final one being our D-charts. In our experience we are witnessing that most of them are born either at the commencement of or end of a

Tatwa and sometimes at the middle-but this should not be taken as occurring always. Sometimes the opposite sex is born at transition times from aroha to Avaroha and vice versa. This will help you to strike a nearer point of birth time as the duration of a Tatwa is long sometimes extending to 30°. In the case of female births it is easy to fix up with the aid of Tatwa only as there are only two female Tatwa that intervene at different periods on like the consecutive ones of male Tatwa. If you are so very particular of striking further rectification under Tatwa only you may go to the Anther Tatwa and Sookshma Tatwa as a male is born in Male Tatwa, Male Anthar Tatwa & Male Sookshma Tatwa. The way of calculating these is just in the way dasha, another dasha and Sookshma dasha *are* done. In this case one full circle of aroha and Avaroha constituting 3 hours duration should be taken as a unit and the sub periods calculated starting from the particular Tatwa noting its aroha and Avaroha. A detailed table in this respect will be published later. But this will not help you much to initially arrive at the result as there are many like sex Tatwa intervening at very short intervals under anther periods. It will only be helpful to check up the correctness of the time once fixed by other methods. So at present it is enough we confine to major Tatwa only. Even here there are some side-trackers who are not aware of the Avaroha cycle and so confine only to Aroha. Don't be misled by them.

Panchanga or Ephemeras Method.

For several reasons Ephemeras method is the best and convenient too as we need all positions in degrees and minutes for our minute calculation. So follow Raphael or Lahiri.

Rasi—Bhava—Cuspal—Charts.

We usually do not recommend prediction direct from the Rasi Chart alone. Anyway if it so necessitates you to do so consider the following elements carefully and diligently as without them you are sure to falter. Though you proceed in the old way of others yet by coupling with some of the rules under our New Techniques you will outshine others though not be cent percent successful in all aspects as Rasi Chart is a heap of many things. What are they ?

1. It has become the habit among the astrologers to consider all things from the Rasi positions alone. Please note it is not correct. This will be correct when the Cuspal and the Bhava charts are alike the Rasi Chart but when any of them differs then comes the trouble. The correct procedure is to consider the ownership of Bhavas from the Cuspal chart i.e., the owner of a house, will be the Lord of the Rasi in which that Bhava cusp (central point) falls. To know in what bhava such lord is, you must locate in the bhava chart. To know in what Rasi he is and what are the aspects and conjunctions with him are to be known from the Rasi chart.

About conjunctions and aspects to be seen in the Rasi chart there is a ticklish point to be specially and carefully noted. A planet may go to a particular bhava as seen in bhava chart but he may not have any connection (aspect or conjunction) with that bhava as aspect and conjunction depends on Rasi chart. As aspect and conjunction have to be decided from Rasi aspect only and not by mere degree difference we have to see if that planet aspects or conjoins the Rasi in which the cusp of the said bhava falls. For example in the case of a Dhanur Lagna chart suppose the 10th cusp falls in Thula. Then lord of 10th bhava becomes Sukra and not Budha as may be visible in the Rasi chart. Again suppose the 10th cusp is 5° Thula and Guru is at 28° Thula then you have to read as follows :-

Guru Lord of Lagna is in 11th bhava and not in 10th for purposes of judging the strength and merits of Lagna bhava. When we discuss the effects of 11th Bhava you should not say that Guru has conjunction with 11th bhava as the cusp of 11th falls in Vrischika and Guru is in Thula not aspecting Vrischika. As 10th cusp falls in Thula where Guru is we can say that Guru conjoins 10th Bhava and so does good to 10th. Thus while reading the effect of Lagna bhava his position in 11th comes while reading the effect of 10th bhava his association with 10th comes. So the two sided effects are as follows:—as lord of 1 in 11 he promotes the effect of Lagna while as mere benefic Guru and also as lord of 1 joining the 10th cusp promotes the 10th Bhava. So he never causes the 11th house. This is how a Bhava phal has to be dissected.

2. Use a Bhava Chart to merely find the position of a planet as either as lord of a bhava in such and such a bhava or to estimate the natural characteristics of a planet being in such and a bhava.

3. From the Rasi chart find the condition, aspect and conjunctions on bhavas and planets. Things to be noted here are whether one is in own. Exaltation or otherwise, Asta or combust or Retrograde etc. in motion as these alter the effects of a planet.

Thus considering all the above 3 charts you have to settle the issue as narrated in detail is our lessons .on Bhava Phal.

CHAPTER IV

Shadbala

In general Astrological books for the evaluation of Shadbala or its kind a lot of rules are enumerated which considered in Toto will merely confuse one. There may be many aspects to be considered but most of them may not act to a perceptible degree. So we have selected only the most effective ones leaving off the rest. Even out of the select few Balas narrated in our text we prefer to confine ourselves to and attach primary importance to--Retrogression, Exaltation, Digbala and Vargottama Balas. These Shadbala speak only the magnitude of effects and never the nature of effects as good or bad. But the one exception is in the case of Swatchetra or own house position when all planets good or bad will always cause good. The rest narrated above merely speak of the magnitude of effects:- good or bad to be judged by other cannons. These play a surer part in our Division Charts than in the Rasi Chart.

CHAPTER V

Yogadhayaya

Other books have gone at length in detailing a long list of Yogas of which many are non-existent in most of the horoscopes and some though existing do not reveal the said effects. Also each type of yoga is said to cause some special results which also do not happen to be so. What we need is the correct mode of dealing and interpretation suited to modern conditions as those such in the text are merely classical appli-cable to ancient times. We have concised the yogadhya and said in a very few words in a nutshell. You may totally forget the names of the several Yogas cited in the books and their effects as matters connoted with the natural characteristics of the planets forming those Yogas are merely shown to come to the scene. In fact it is not only the natural characters that will be revealed but also the more important Functional traits arising out of ownership of bhavas. So while judging the yoga effects both the natural and functional traits have to be considered and then read out.

You may try to forget the names of the several Yogas as those names are given only to remember the classifications. We are analyzing below how these Yogas are formed:-

1. One set of Yogas are formed by single planet being situated in Kendra identical with own house. What is the specialty here. This is one of the tests of weighing a planets trait discussed under Bhava Phal. These fall under Ruchaka Bhadra etc. Yogas. In fact it is not only Kendra but all good houses do cause similar Yogas.

2. Next set of Yogas are formed be planets being on either side of Lagna. Sun or Moon and these are Vasi, Vesi etc. types.

3. Next set is caused by more than one planet say in sets of two each being posited on either side of Lagna. Sun or moon. In fact this is just the singled out type of the General Sankya Yogas narrated in Jaimini. According to Sankhya Yoga planets at equal distances from Lagna form this yoga— good or bad -depending upon the natural benefic or malefic nature. We always prefer Yogas formed round about Lagna being more powerful than the one round about Sun or Moon.

4. Parivarthana Yoga Caused by two planets :

This is caused by the exchange of houses between any two planets. Here you should not go on the blind old method of counting from Lagna in the Rasi Chart. What actually are the cusps situated in the two Rasis occupied by the two planets are to be noted and it is only with respect of these that there is exchange and no further. For example in a nativity where the 10th cusp falls in Vrishbha and 9th in Meena (say .for one born in Kataka Lagna by the altered positions of cusps) you can say that there is exchange between the lords of 9 and 10 when Sukra is in Meena and Guru in Vrishbha. As Guru and Sukra own other house too you should not say that the lord: of these houses also are exchanged. It is only the cusps falling in the Rasis occupied by the planets that get exchanged and not the other set. Sometimes if more than one cusp falls in any of the Rasis then you should say that all those bhavas get exchanged.

5. Gajakesari Yoga :

This is formed by Guru being in Kendra from Moon. This requires further amendment. This sort of mutual position between Guru and Moon is seen in plenty of charts but they will not be enjoying all the effects ascribed to this yoga. As this and any other yoga is caused by some planets it is very necessary that those planets must be well situated as judged under the theory of Bhava Phal. What is the good by such a guru being lord of Dustanas or being Asta or in Dustanas or in Neecha etc. Thus without judging the strength (by Shadbala) and good or bad of the planets (by Bhava Phala) never conclude the effects of any yoga.

6. Viparitha Raja Yoga :

The name of this yoga we wish to slightly modify as Viparitha Yoga. This is caused by the exchange between the lords of Dustanas or by the lord of a Dustanas being in Dustanas. Unless those exchanged planets are connected with Sun how can there be Raja Yoga and so we are correcting its name as applicable in all cases. By Raja Yoga is meant in general sense that one will have connection with Government in some form from peon to Maharaja depending upon the strength of the planet or planets that cause it. The differences between right royal yoga and Viparitha yoga is that in the former case we expect some good thing to come and they really come but in the latter case things we expect may not come but sudden and surprise events will fall on us like the bolt from the Blue. The one attained in the former way is by easy process without any obstruction on its way while that of the latter is after many obstacles and hindrances, and sometimes undeservedly.

NOTE— 1. All these act with cent percent success in our D–charts though some of them may fail in the Rasi chart.

NOTE-2. Yogas caused by single planet are seen during that planet's Dasha while those caused by two are seen during one's Dasha and another's Bhukti. Those caused by 3 in one's Dasha, another's Bhukti and the third's Anthar.

CHAPTER VI

This chapter on Female horoscopy is very simple and explained in full in the text itself which does not need any more revision.

CHAPTER   VII

This is the last chapter in this part and is it the most important subject, dealt with at last. For proper handling of the Rasi chart alone you must master this chapter well. In short space and time as here it is not possible to reopen the entire theme in Toto. So in a nutshell we will only impress on the most important aspects you must lay your hand on for correct handling and judgment.

1.   In judging the effects of a Bhava say 4th bhava first find out in what Rasi the cusp of the 4th falls. Say it falls in Mesha for a Dhanur Lagna birth. Then Mesha is the guiding factor for your further dissection about the 4th bhava and not Meena though counted from Dhanush falls in 4th. This you will note from the Cuspal chart.

2.   Then find out what are all the planets situated in this Rasi (Mesha) and aspecting this Rasi. This should be judged from the Rasi chart only as aspects and conjunctions will have to be considered from the Rasi chart alone. Considering both the natural and functional qualities of the planets judge the bhava in question.

3.   Next find where the lord of this bhava i.e., Kuja in this case is situated. This has to be judged from the Bhava Chart. Here read only the effects of the bhavas he owns and not where he is. Suppose he is in 7th bhava it is only the good of the 4th bhava that has to be read and nothing about the 7th bhava.

4.   Next consider the star in which the planet is situated and modify the effect of this owned bhava (4th) nothing the nature of the lord of the star (Lord of the star is the dasha lord).

101

5.   As a final measure judge from the stellar lord of the cusp of this bhava. Suppose the 4th cusp falls in Bharani its lord is Sukra and see how Sukra is situated in the chart and then conclude the merits of this bhava (4th). This is what is known as Sookshma lord of a bhava while the Rasi lord of the cusp is Sthoola lord.

6.   Never forget to consider the Birth Yogi, Birth Avayogi, Duplicate yogi and Zero Rasis along with the above criteria.

In this way judge Bhava by Bhava individually and finally you will be able to judge the entire horn. scope well.

Tests of Progeny

This is wholly true. Unless you apply these preliminary tests never venture to predict the aspect of issues in a horoscope.

So sir, we have not only coached you up in these preliminaries to the best of our ability and also refreshed your memory to the most salient features in this lesson. It is now up to you to have a thorough study or our text and all the lessons so far sent and digest the foundation very well so that you may find it easy to follow our ensuing lessons getting into the actual arena of predictive astrology in a unique manner founded by us and us alone for the first time in the history of this noble Vedic science and that is our Masterly New Techniques of my Guru commended by the leading 'Hindu' in their Review of 29-9-1963 who have opined it to be the first of its kind in that it is an exhaustive and self-sufficient one worth being followed by even the practicing astrologers. Of late it is being appreciated all over the world.

With this we close this lesson wishing you the full enlightenment on the subjects so far taught.

102

LESSON No. 13

We have hitherto discussed in all our previous lessons about the fundamentals on which the predictive astrology rests. We hope you have mastered them all Now we step into the real predictive side which is the crux of the science and of which you have been anxiously awaiting to know. In the following lessons our method of approach is a little different from the usual classic method but not entirely different from it. You may call our method as one polished over the rusted one.

As we have already pointed out it is the sheer neglect to consider the five Angas of a Panchanga (Thithi, Weekday, Star, Yoga & Karana) that falsifies the correct reading of a horoscope. In this lesson we confine only to two of them the rest being taken up later, and they are Thithi and Yoga. You know already that these elements are got by the difference and sum of the longitudes of Sun and Moon and you also know that for a person in a particular Thithi some of the Rasis become Zero Rasis (Vide Ch. I Pt. I PP. 9 of our text). There you see at least two Rasis going bad except in the case of Chaturdashi when four Rasis become zero Rasis and in the case of Full Moon and New Moon days there will be none. What are its effects By Zero Rasi is meant that sign is burnt (Dagdha Rasi in Sanskrit) or in other words that it becomes depressed. Consequently the Bhava that is situated in that sign becomes negative in character-becomes opposite of its real character i.e. a good bhava becomes bad and a bad bhava becomes good. So it follows that a planet in such a zero Rasi becomes ineffective or depressed, just as in Neecha house. In the same manner the lord of such a zero Rasi has to be viewed as being a negative planet just as one owning Dustanas. If such a lord is powerfully and auspiciously situated in good bhavas he gives inauspicious results with respect to the Zero Rasi Bhava he owns. But please note that this sort of a negative effect will be only with respect to the zero Rasi bhava he owns and not of the other non-zero Rasi bhava he may also own. For example for one born in Thule Lagna on a Padyami Sani becomes the lord of Zero Rasi fourth Bhava and of non-Zero Rasi fifth bhava. Suppose that Sani is in Mithuna then lord of negative 4th bhava in 9th he causes distress to the 4th bhava while as lord or positive 5th causes good in matters of the connotations of the 5th bhava, If on the other hand that Sani is in 8th bhava then as lord of negative 4th in 8th becomes good while as lord of positive 5th in 8th goes bad. Thus you must discriminate when a planet owns two types of Rasis (Zero and Non-Zero Rasis) simultaneously.

Please read thoroughly Astrology Book of our text dealing on Thithi and Yoga. In a nutshell we can tell you that the entire mode of judgment is centered round the theory of Double Positives, and Double Negatives making one positive and Positive versus Negative making negative. It may be understood this way. Any functional malefic (owner of Dustanas) or the lord of zero Rasi (negative planets should not become powerful (by situation in good bhavas, in exaltation or retrogression) but should become powerless (by being in Dustanas. Zero Rasis, Debilitation and set by Sun) to give good effects. It is just like a mischief-monger being allowed a free hand to do as he likes and thus allowing him to go strong to do the mischief. If on the contrary the negative planet we cited above be weakened he gives good effects of the bhavas he owns as otherwise bad only results. If you understand this basic principle you will have understood all the theories enunciated n this chapter.

This is all the sum and substance of the mode of using Zero Rasis which are elaborately discussed in our text. So by the mere sight of a Rasi becoming a zero Rasi or for the reason of a planet being located in a Zero-Rasi or Dustanas do not suddenly jump 'in to any sort of conclusion until all the further tests are completed. You must further see if there will be no further modification such as that Zero Rasi lord being in another zero Rasi, in debilitation, in Dustanas or set when he will give good results. Similarly of the planet in that zero Rasi see if he is retrograde he owns a zero Rasi or Dustanas. In this way by further scrutiny only you must come to final conclusions coupling the nature of the star in which it is situated. Never jump in to sudden conclusions at first sight as you know the proverb 'Appearances are deceptive'. This is all the sum and substance of the effect of Thithi you have to note.

Next we go to Yoga-a very important organ molding the real worth of a horoscope. You know that Yoga is got by adding the longitudes of Sun and Moon and you have also known how to fix its point in the zodiac. Now we shall discuss its effects. When once you fix up the exact point of Birth Yoga in a horoscope you treat that point as if it is a Cuspal point as in the case of a bhava cusp. Then you find the Sookshma and Sthoola lords of this cusp. You know that the Sookshma lord is the Udu dasha lord of the star occupied by the yoga point and the Sthoola lord is the lord of the Rasi in which this yoga point falls. In other words the Sookshma lord is called the original birth yogi while the Sthoola lord is the Duplicate Yogi Though both of them play parts in a horoscope yet the original birth yogi plays a major role. For example if the yoga point be 3°40' Meena which corresponds to Uttarabhadra star Sani becomes the Birth Yogi and the lord of Meena where the yoga point falls Guru becomes the Duplicate Yogi. Most of the astrologers of the day are not hitherto aware of the importance of this Thithi and Yoga effects, which was hitherto kept secret which is *now* revealed out by us for the benefit of mankind and it is these that play tremendous role in the study of horoscope. Thus if you neglect these two important elements you will be nowhere. How wonderfully these elements work will be discussed in detail in our next lesson.

Simultaneously with the yoga planet there is what is known as Avayogi planet which acts in the opposite manner of the Yoga planet. How to locate this Avayogi. The Avayogi is the sixth Dasha lord from the Birth Yogi. On page 2 Ch. I Pt. II against each Yogi its corresponding Avayogi is given. From this table you may find out the Avayogi corresponding to the Birth Yogi. This applies only in the case of original yogi and not

107

of the Duplicate yogi. Thus there are two
yogis and one Avayogi. Further you notice
that there are two sets of planets falling into
three groups and they are—

Ketu-Moon-Guru (I set),

Sukra-Kuja-Sani (/II set),

Sun-Rahu-Budha (III set).

Planets of one group are mutual friends in
that if any one of them becomes a yogi the others
become like yogis by dint of group friendship. In
fact these Avayogi are also termed as enemies of
those yoga planets. Thus under this system only
one planet is the enemy of the other. But this
theory of single sided enmity will have to be used
only when we deal with the Novel Dasha chart and
at all other times the general friendship etc.
described under Parashara to be followed as
described in our lesson I.

Await our next important lesson on the handling of
Thithi and Yoga effects.

LESSON No. 14

Subjects dealt with in this lesson refer to Chapter I Part II of our Text

In our previous lesson no. 13 we have dealt with the effect of Thithi and Yoga on the life of an individual and the mode of fixing up of these points in the zodiac. In this lesson we discuss at length as to how these two elements act in molding the destiny of a person. As stated and even warned before, predictions made without considering these factors will many a time go wrong. Hence you should never forget to consider these important elements along with the usual cannons of judgment.

How to Read then mixing up these additional factors:

To fully grasp this theory we have classified the Planets, Bhavas, Rasis, and Stars as Positive and Negative. This must be thoroughly understood from our texts wherein they are clearly and exhaustively explained.

As per this classification a planet attaining Positive character and situated in Positive Bhava gives good of the effects of the Bhava it owns and of its natural qualities too. Thus, its situation in Bhava judged from the Bhava Chart is of paramount importance in the determination of good or bad of the effects of the Bhava it owns and of its natural characteristics. Then you may consider the Rasi effects which only speak of the magnitude of the effects judged before. Suppose a Positive Planet (one owning Positive Bhava) is in Positive Bhava (seen from the Bhava Chart) but in Negative Rasi (in Rasi Chart)• Then the planet gives no doubt good

of the good Bhava it owns but of a lower magnitude in effect-
On the other hand if that positive planet is in a negative
Bhava it gives bad for the good Bhava it owns, if such a
planet be posited in a positive Rasi especially in own house he
tries to maintain his status-quo and tries to give soma good at least
for the good Bhava it owns.

For Example.

For Thula Lana, Sani, lord of 4 and 5 (positive
planet) in Lagna Bhava (Bhava chart) being situated in
Thula gives good being all positive (good for 4th Bhava
and 5th Bhava effects and his Karakatwas as service,
longevity etc. Suppose he is in Mesha and in 7th Bhava
(as per Bhava chart). As a positive planet in positive
Bhava he no doubt gives good of the effects of 4th and
5th Bhavas and of his Karakatwas too hut being in
Neecha sign (negative Rasi) the good is reduced in
magnitude and yet some good stands in the end though
little. If on the other hand if he were in Means as 6th Bhava
he would not cause good for the said 4th and 5th Bhavas as
the positive planet will have gone to a negative Bhava.
Though Meena is a friendly sign (Positive Rasi) by reason
of being in Negative Bhava bad only predominates. Thus
you see that good or bad it is the Bhavaic Position in the
Bhava Chart that is more important than its position in the Rasi Chart.

Similarly, a negative planet in a Positive Bhava gives
bad effect with respect to that Negative Bhava it owns. If such a
Negative Planet be situated in a Negative

Bhava it gives excellent resuⁱt with respect to the Bhava
it owns and this is termed as Viparitha Raja Yoga'—a yoga
which comes all in a sudden unexpectedly after some bad.

Suppose this Negative Planet being in Positive Bhava is in a Positive Rasi then it gives worst results with respect to the Bhava it owns. Please note that always we speak of the Bhava it owns If such a planet be in a negative Rasi then it gets slightly modified and tries to cause lesser evil—at any rate the net result is had.

You know that a Positive Planet in Positive Bhava or a Negative Planet in a Negative Bhava gives always good. This is nothing but the theory of double positives or double negatives make one positive. Likewise a Positive Planet in a negative Bhava or a negative Planet in a Positive Bhava causes bad by the theory of positive into negative gives negative.

Next in importance come the starry effects. In this connection we wish to impress upon our students that in very old days they wholly based the entire predictions on the Stars only and Rasis are of recent use in fact when we deal with Udu Dasha we should deal all things and measure all effects only as per stars. By the situation of a planet in a particular star it imbibes the character of the star in which it is posited. Again the same theory of positive versus negative or positive as we discussed above holds good even here. Instead of Bhava we take up here the nature of the star in which the planet is posited. Like a positive planet being in positive or negative star. Note that the positive or the negative nature of a star is judged by the positive or negative nature of the star lord. Though in general this modified character of a planet in a star is seen through its Dasha yet its striking effects are seen chiefly in that quarter of the Dasha of the planet corresponding to the star pads. Suppose a planet is situated in 3rd pads of a star then this modification has to be read out, in the 3rd quarter of the

Dasha of the planet. Here one special thing has to be noticed. You must first assess the merit of the star lord as per the above principles of positive and negative characters and then read the effect of the stellar position. For. For a positive planet in the star of a negative planet may at the outset show to be bad but if t^hat negative stellar lord is further in a negative position as he thus attains positive trait good only should be predicated. So the position and association and aspect on this stellar lord will have to be considered. Though all these appear to be of multiple alternatives a little practice will make you perfect in this art. As stated above this modified effect due to the stellar position will have to be read out only during the Paid quarter of the Dasha and at other periods of the Dasha the original effects of the said Dasha lord will happen.

Example

In the above cited example suppose Sani is in Punarvasu 2nd pada and in 9th Bhava too, then the effects are:—

During the entire period of this Sani Dasha of 19 years Sani gives good of the effects of the 4th and 5th Bhavas he owns as he is in Positive Bhava but with the modification that during its second quarter of its Dasha i.e., from $4_3^{/4}$ years to 91/2 years of this Sani Dasha it gives the effect of Guru As this Guru is the lord of 3 and 6 Bhavas (Negative Bhavas) by the theory of positive Sani getting into Negative Guru the effects of the 4th and 5th Bhavas get modified during this period as follows. At the outset we have to read as bad. Then you should look to the condition and situation and aspects of this Guru. If this Guru is in a negative house then you have to speak of good only later. But if that Guru is in a positive Bhava then bad only predominates. Thus the position. -aspect and conjunction of the stellar lord have all to be assessed before pronouncing the effect of the Dasha lord. Please remember once again that the ownership has to be judged from the Cuspal Chart, Bhava position from the Bhava Chart. and Aspects and Conjunctions and condition from the Rasi Chart. One with Guru or Birth Yogi or aspected by them will prove to be good while with or aspected by malefic or birth Avayogi do badly. A planet with or aspected by his enemy will obstruct the happening of the good however powerful the planet be. So inimical combinations or aspects are not productive of good Here one thing has to be noted. Between any two planets one may be the enemy of the other while the other may not be the enemy of the former In the case of mutual inimical planets the Dashas of either get spoiled while in the case of partial ones the effects will be as follows. The Dasha of a planet connected with his enemy gets spoiled and not the Dasha of the other who may not be the enemy of the former. For example if Guru and Budha are conjoined or mutually aspecting each other it is the Dasha of Guru that may get afflicted as he is with his enemy Budha, but the Dasha of Budha will go good as Guru is not his enemy. In the case of Ravi Sani combination the Dashas of both go bad as they are mutual enemies- This is how the inimical effects to be judged. It is the habit of the old school of as! Astrologers to speak of total bad to both the dashes once such a combination exists.

Cumulative effects and Cancellative effects.

Here you have to know as to what would happen if a planet has more than one character (positive or negative) to start with and similarly more than one at ending The start is at the point of ownership and the end is at the situation in Bhava. Rasi and Star. For example one owning Dustanas which is also a Zero Rasi and himself a Birth Avayogi (all

114

negatives combined at the start) becomes more negative and not cancellative. Similarly, one posited in Dustanas which is a zero Rasi, Neecha Rasi and also set by combustion with Sun will all increase the negative character and again becomes cumulative and not cancellative as before. Such added effects of the same nature are called cumulative effects meaning enhanced nature of a single nature as in this case of negative and in some case may likewise be positive also if all of them are of positive characters to start with or end with. If a positive planet to start with at the beginning is in a negative Bhava, Rasi or Star then only the cancellative effect comes in. This must be clearly understood as otherwise you are liable to commit errors in judgment,

Some Special Points to be noted.

1. A Birth Yogi always gives Yoga (Material Prospects) in whatever position or condition he be—his grade of effect may vary but the ultimate good is always there· By Yoga is meant only material prosperity and nothing beyond it. One may die even during the period of a Yoga Planet and then you may ask why he died during yoga time.

2. Likewise whoever is in the star of birth yogi gives yoga.

3. A. planet with or aspected by Birth Yogi causes yoga.

4. What is said of the Birth Yogi equally applies to Duplicate Yogi.

115

5.   As we have spoken of good yoga with respect to Birth Yogi so Avayoga (bad) will happen in the case of Birth Avayogi acting as above.

6.   If in a horoscope the birth Avayogi becomes the Duplicate Yogi then there will not be downfall in his life.

7.   The Birth Yogi or the lord of Lagna though owns a zero Rasi still remains good.

8.   A planet in the star of Birth Yogi or Duplicate Yogi is good and excellent if he be in the actual birth yoga star. Similarly bad to be read out in the case of Avayogi stars.

9.   Birth Yogi and Birth Avayogi combined or mutually aspecting makes the yogi go a little bad during his period while the Avayogi period gets on slightly better just like the wicked fellow in the company of the good tries to imbibe some of the good traits of the other while the good fellow contracts some of the bad traits of the other.

10. For full benefic effects the Birth **Yogi** should become strong and the Birth Avayogi must become weak.

11. If Lagna falls in the star of the Birth Yogi he will be prosperous froth birth to the end of life and certainly after the Dasha of the lord of Lagna or the lord of the Yoga star Dasha. Likewise if it is in the star of Birth Avayogi bad only emanates.

12. Whichever Bhava lord (Sthoola or Sookshma) becomes the Birth Yogi it may be said that the native will become prosperous with it. Suppose it is in the case of 7th Bhava then his prosperity increases after marriage and he will have some lucky chances through the wife or her side. If it is the case with 5th Bhava then at the birth of each child some promotion is awaited and so on. Likewise if they are connected with Avayogi read bad effects only.

13. If the lord of Lagna is Birth Yogi or Birth Avayogi read as in Para 11 above.

Thus in this lesson we have touched upon the most salient features of the effects of Thithi and Yoga which are not known to the public so far. For further and detailed study please read our text carefully and note the examples given there In this lesson there is the crux of the subject that is of primary importance which if ignored will certainly upset the entire readings. So it is now your responsibility Jo master this aspect and be prepared to go through our next lesson on Division Charts—our Masterpiece.

LESSION – 15

In this lesson we deal with the Division Charts—our Masterpiece—the outcome of long research and that by practical application to known horoscopes the, theory of which is not known to any one so far, nor the mode of using them. You do not have all the Division Charts described in the classical texts of which Brihat Parashara Hora Shastra is the standard authority. Even there all the Division Charts that we have illustrated are not to be found. Luckily we have been able to find out some of the rare divisions from some of the rare Manuscripts and Editions. Some books describe the Modus Operandi in the method of erection of these charts in all sorts of ways. But alas none of them seems to have taken the trouble to find out by practical verification as to which would suit best in all cases. They would have attempted if they were aware of the proper rules of judgment as there are again distorted modes of judgments described by authors in their own ways. Based on the principles narrated by the classical authors coupled with research by practical applications to known horoscopes we have been able to formulate genuine Principles of judgment of these charts which, when applied by yourselves will convince you of the truth in our statement.

How to erect these Division Charts.

Some modern authors have suggested some modes to erect them but none of them fit into practical tests avid so we have, basing on the ancient principles narrated in rare manuscripts, given complete rules on this subject in pages 49 to

56 of Pt. II of our text. As they are self-explanatory please study them and learn. If at any stage you feel doubtful on any point please verify from Chart 1 given in the Appendix Booklet which is no other than that of the author himself in which case all the D-charts are worked out in full. As they are based on defined mathematics we do not wish to repeat them once again here and thus waste time and space.

Note some of the peculiarities here.

1.   Hereafter we call the Division Charts by their respective numbers as D-1 (Rasi Chart), D-2 D-3 etc. as per the number of the Vargas or Divisions they bear as they are no other than the Varga-Kundali used for the estimation of the total strength of a planet as Shadvergabala etc.

As in D-2 (Hera) all planets and Lagna will be in only two signs (Kataka or Simha) no useful purpose would be served by considering this chart and also it has been practically verified to be incorrect for application. So we leave it.

2.   Under D-3 (Drekkana) there are two modes of working Take only Varahamihira's version that is followed by us,

3.   Regarding certain D-charts some have twisted the mode of working. We only regret to state that they have been mere copyists of some other author without taking the trouble of verifying the truth of them by practical applications. Alas, unless they know how to use them how can they verify. We have after mature thought, experience and practical applications to very many known horoscopes come to certain candid conclusions which agree with the principles

119

enunciated in rare Manuscripts.

4.   D-30. Texts merely mention the
Trimshamsha Lords and do not assign places in the Rasi
Chart. Unless any factor, if at all interferes with one's life, is
brought inside the chart and a definite place assigned to it does
not help us to make use of it. We have therefore fixed up the chart
for this D-30 also.

5.   On page 57 of Pt II of our text a Table is given
to aid the finding of the Ameba parts of some of the D-
charts whose working involves fractions. Those charts
consisting of whole parts may be worked out mentally.
Please study how to use this table from our text. You may also do
this from first principles.

6.   We call the Original Rasi chart as D-1 and the
rest as D-2, D-3 etc. after their number of divisions.

7.   Please understand that these Division Charts
are no other than the several Vargas Kundali is that you are
aware of in general Astrology as helping the evaluation of
Shadvergabala,   Sapthavargajabala,   Dashavargajabala,
Dwadashavargajabala etc. used in estimating the total strength
of a planet. This sort of finding the total strength of a planet is
not only cumbersome but also serves no real purpose. For, as
you may see later, a planet possessing strength in any one D—
chart will only augment the effects pertaining to that D—chart
and no further. Similarly weak and badly placed planet in a D—
chart will give less and bad effects too of that portfolio of that
D--chart. Under such circumstances if you club more than one
character together by considering the total strength arising
from different D-charts and judge the effect by its combined
strength you get only wrong conclusion. For example if Guru

has 10% strength in D-24 (Education Chart) while 100% strength in D-10 (Profession) the correct way of judgment would be to ascribe 10% results for education and 100% for profession and not say that striking an average that both these effects will have 55% results which will be far from truth. Of course the sum total strength of the planets will help us to estimate the comparative strengths of planets when comparative results are to be judged but in the evaluation of single sided affects this are not needed.

Now that you have understood in clear terms about this point please do not err at least hereafter by placing undue importance on the sum total strength of planets.

How to judge the Division Charts

Some authors have again misled us and created confusion in the mode of valuing the effects of a D-chart. They are again to be called side-trackers from the genuine way. We are after mature thought, research and practical application convinced that our mode of application is correct one and that we have placed before you-Please note that whenever you deal with a D-chart you must confine only to its particular effects and no further. Thus there is a special advantage had from these D-charts which speak of only one aspect in life.

How to ascertain the good or bad of a planet in a D-chart.

This is very simple and can be said in a line which is so sure and as true without exceptions as stated below. If a planet whether natural benefic or malefic, is situated in good positions from that D-Lagna then it always gives good results. If on the other hand it is in Dustanas (3-6-8-12) it gives bad effects. Is not this one rule so simple definite and certain. After knowing the good or bad that may be caused by a planet in a D-chart you will have to next assess the magnitude of its effects by the Shadbala of the planet in relation to that D-chart. Beware it is Shadbala and not Shadvergabala that we want.

Are there any exceptions to this rule.

Yes, every rule has an exception and so this also and they are-

1. It is said in general astrology that all planets in 6th house become bad but exception is given to Kuja for the reason that though in 6 be aspects Lagna. You know for all good a planet should aspect Lagna. On this theory Kuja becomes good in all D-charts though in 6th from that D-Lagna.

2. Similarly it is cited in general astrology that all Planets in 8th house cause worst results but exception is in respect of Budha in matters of causing sudden wealth by will or legacy or by the death of someone. But the same Budha may cause distress in all other matters. Applying this rule to D-charts Budha in 8 in D-4 (Assets) is good while in all the other charts he causes bad.

3. White discussing the effects of Karaka-Bhava it is said that Sani in 8 is an exception who causes high longevity. In the language of our D-chart Sani in Lagna in D-8 causes high life.

4. In a D-chart a planet in Dustanas may not harm much if he occupies his own house but will, nevertheless be a little bad.

5. Finally one special thing has to be noticed which if neglected will give you wrong reading. If a planet is in

Dustanas in a D-chart do n°t suddenly jump into conclusion that he causes bad. See if a natural benefic is at the same distance on the other side of the Lagna thus forming Sankhya Yoga Suppose in D-9 Guru is in 8th and his Dasha running. Your immediate impression is that Guru Dasha is bad to wife (D-9 is wife chart). If suppose Sukra is in 6th thus forming Sankya Yoga with the 8th placed Guru, then the bad of Guru is remedied. Thus all the Yogas cited by us under Yogadhayaya should be applied to all D-charts.

6. More than all it should be specially noted that the D-charts effects being one of periodical nature unlike the permanent effects of a Rasi chart whenever you read effects of a D-chart the current Dasha Bhukti should be considered.

Above are cited the mode of ascertaining the good or bad effects of a planet arising out of /its position in the D-chart with respect to its Lagna. The next question is to assess the magnitude of its effects which will be explained in our next lesson.

As lessons hereafter are very important for predictions follow carefully and digest them well. Await our next lesson

LESSON No. 16

In the previous lesson we have discussed the formation, elucidation of the good and bad effects -f a Division Chart and the fallacy in basing predictions on the sum-total strength of a planet (Shadvergabala etc.). In this lesson we consider the mode of establishing the quantum or magnitude of those effects from a D-chart. To do this you must be fully conversant with Ch. IV Pt I (on Shadbala) of our text You may be aware of the slippery say of some of the old school of astrologers the from the horoscope of a person it is not possible to assess the actual grade of one an

So they suggest considering the family in which one 'is born. One born in King's family will become a king if endowed with Raja Yoga in his horoscope while the same yoga in one belonging to ordinary family gives only high position in life. Though this is partially true it is not wholly so as may be judged from the present day conditions when there is no king on earth and we see even ordinaries getting the highest seats in government. Thus birth alone should not be the guiding factor in striking the grade. The actual strength of a planet will cause the actual grade.

How then to judge the magnitude of effete.

It is simple rule of commonsense that the magnitude of effect depends upon the strength of a planet. Apply this simple rule to the particular D-chart. Once again note that we do not want the sum-total strength of a planet in all the divisions. What we need is to find out its strength by Shadbala in a particular D-chart only. To do this we refer you to our

Chapter on Shadbala (Ch. IV Pt. I). Even here we recommend you to limit yourself to the most important ones ignoring the rest as they do not count very much in the total effect We say this by practical experience. Remember always that, whatever we say candidly and emphatically are the outcome of our long research by application to known horoscopes and not guesswork as most of the present day astrologers do. For purposes of revision we once again quote the most important items of strengths to be noted.

1. A planet in own or exaltation house is powerful but in Neecha weak. Here we omit the friend's house, neutral's house and even enemy's house as they will after all modify the effect to a very small extent.

2. One in Vargottama, with Digbala or Retrogression becomes powerful.

P.S.--By Vargottama it is generally construed as being in the same Rasi both in Rasi and Navamsha charts. This is not all. A planet in the same sign both in the Rasi Chart (D-1) and in any of the D-charts will have Vargottama position so far as that D-chart is concerned.

Likewise the Digbala also has to be reckoned in the D-chart from its position from the D-Lagna.

3. A planet that is set (Astha) in the Rasi chart becomes weak in all the D-charts.

P.S.--Retrogression and combustion *(set)* have to be found from the Rasi Chart only as they exist in the heavens.

125

4.   Do not reckon Zero Rasis in the 'D-char is. This should be looked into only in the Rasi chart. But Birth Yogi. Duplicate Yogi and Avayogi should be considered in all the D-charts. One thing to be specially noted in this connection is that while considering the Yogi etc., in a D-chart so far as it relates to the person of the horoscope those Yogas act and not in respect of others. , For example in D-9 D-10, D-1 I, D-16, D-24, and D-4 the yoga planets work as these charts relate to the prosperity and happiness to self. But in D-12 this yoga planet should not be used while reading the effects of parents. In D-6 and D-8 these have no special significance. Yoga means only material good and nothing beyond that. Health, accidents etc. do not come under this category. So even a yoga planet in D-8 may kill the native.

5.   So far we confined ourselves to the use of noting the Shadbala in the D-chart only. We go a step further and add on the Shadbala strength possessed by the planet in the Rasi Chart. A planet may occupy a very ordinary but favorable position from its Lagna but if that planet possesses any of the Shadbala in the Rasi Chart though this planet be in bad positions in ilia Rasi Chart so far as the effects of this D-chart is concerned it gives high good effects. We will elaborate this as follows:-

a) If the planet in D-1 (Rasi Chart) occupies its own or exaltation sign or retrograde or with Digbala then even an ordinary good position from D-Lagna in the D-chart will cause higher grade effects. This will be so even if that planet in D- I is in Dustanas. Then the question may arise as to what aspect of life he shows bad being in Dustanas in Rasi chart. The answer is simple it may be with respect to some other aspect where it is badly situated in a D-chart. But if a planet however powerfully and favorably situated in D-1 is situated in Dustanas from D-Lagna it only gives bad effects with respect of the effects of this D-chart. Thus you see that a mere good position from D-Lagna is by itself enough to conclude good effect with respect to that D-chart and the positions in D-1 do not always and in all cases reveal the actuality. Similarly a planet that is Neecha or in Dustanas in D-1 will be good if in the concerned D-chart it is in good positions from D-Lagna and may even cause high grade effects if endowed with Shadbala in the said D-chart. This is how you have to make use of the strength of a planet in a Rasi chart.

b)   A strongly posited planet in Rasi chart if in a favorable position from that D-Lagna and also possessed of strength in that D-chart then where is the doubt for high grade good effects of that D-chart. Thus if two or more strengths exist both in the Rasi and D-chart then highest grade effects with respect to that chart will happen. This is how we have to predict the magnitude of effect in any walk of life and so it is not necessary that one should be born in a Royal family only to enjoy the highest order of life.

c)   Consider all the Yogas cited in our text— vide chapter V Pt. I. If any of the said Yogas exist in a D-chart then during their periods it causes good effect of that D-chart. For example if Guru is in 12th in D-10 chart the conclusion is that during Guru Dasha the profession fails. But if Sukra the other Benefic be in the 2nd place this bad of Guru gets modified and only good will prevail during the same Guru Dasha as Shubha Sankhya Yoga is formed by Guru and Sukra being at equidistance on either side of this Lagna. But

127

a similar yoga existing in the Rasi Chart mayor may not act and even if it acts we will not be in a position to say for what aspect in life this yoga operates while read out from the D-charts we can definitely say in what way it acts. So you see how wonderfully these Yogas and Shadbala work in a D-chart. All the sayings of our Rishis and Daivaganas are true but we are lacking in the mode of application of these sayings. It is only after applying this way that we were led to respect Parashara and Varahamihira. If a planet is Neecha in a D-chart see if there is neecha-bhanga. Then the bad turn out turn-out to be good. The following are the tests for

If a planet is Neecha in a D-chart see if there is neecha-bhanga. Then the bad turn out turn- out to be good. The following are the tests for neechabhanga:-

1. The lord of the Rasi occupied by the Neecha planet being strong by retrogression, exaltation, or in Kendras from Lagna.

2. The lord of its exaltation sign similarly situated causes neechabhanga

3. The Neecha planet itself being retrograde.

4. If the Neecha planet is either conjoined or aspected by the lord of the Rasi occupied by the Neecha planet.

5. Similarly see the position of "the lord of the star in which the Neecha planet is situated.

P.S. - It is only when the Neecha planet is in good positions from Lagna that the tests of Neecha Bhanga are to be applied as that placed in Dustanas will never regain good effects even if there be Neecha Bhanga for the very reason that a planet in Dustanas in a D-chart is of no avail.

e. No lordship of planets should be considered in a D-chart except the lord of D-Lagna who alone plays some important part. Many a, time the good or bad will commence from the period of the Dasha of the lord of that D-Lagna with respect to that connotation, provided that dasha comes in right and proper age of enjoy ability.

f. All aspects and conjunctions in a D-chart are to be read out considering only the natural benefit or malefic character of the planets.

g. The above are the special points narrated by' us. Along with these please go through all the 37 Rules stated in pp. 59 to 66 Ch. II Pt. II of our text.

h. In judging a D-chart you should not look to the total effects of all planets as the planets give their effects during their periods only. So during the periods of particular planets you must look to them only as Dasha and Bhukti lords and none else. All summary life aspects of permanent nature should be read out from the Rasi chart only while the periodical happenings are to be seen from the D-charts.

Thus we stop here for the present. See our next lesson containing some pointed effects in particular charts.

.Lesson No.17

As stated in our previous lesson No. 16 the present lesson is more or less a revision of the previous one with only the salient features being discussed for recollection. Here we may mention some of the most salient points leaving the rest to be studied from our text.

Reading of D-3. (Co-borns)

In this chart we have to read the effects of both the elder and younger co-horns (brother or sister), uterine or step ones. There may be periods when some of them only may yet affected while others may not. Then how to read such a result.

This should be judged from the Dasha and Bhukti lords for the time. If the Dasha ford.is well placed from this Lagna then partial bad to some only will be felt during the period of the Bhukti lord who is in Dustanas from this Lagna or from the Dasha Lord. But if the Dasha Lord is himself placed in Dustanas then bad for all to be predicted. This is how it should be read out. Applying the theory of 'Karako Bhava Nashaya a' we have said that karaka in Lagna of the concerned D-charts will afflict the particular aspect especially during its period. As this chart includes both younger and elder Coborns the way to discriminate the two is by the respective karakas. If Guru the karaka of elder coborn is in D-3 Lagna then bad to elder coborn is to be read out especially during the period of Guru while Kuja the karaka of younger coborn in D-3 Lagna will harm the younger coborn specially during Kuja period. Similarly such discriminations are to be made with respect to parents when we deal with D-12. For father consider both Sun and Sani and for mother Moon and Sukra. Note that of the two karakas if one stands for yoga the other represents health and longevity. By experience we see that for those born in day time Sun represents health and longevity of father while Moon stands for yoga of mother. For night births Sun stands for father's yoga while Moon represents longevity of mother. Only those aspects that are actually denoted will suffer by reason of Karaka-Bhava.

A similar application should be made in the case of all other D-charts which represent more than one person such as D-7, D-9, and D-12 where multiple children, wife and parents (both uterine and adoptive) are concerned.

Reading of D-8.

In the fixing up of maraka as per this chart sometimes many planets may indicate maraka and thus create doubt In such cases scrutiny of the Rasi Chart will help you sometimes. More than all you must judge the span of life from the. Rasi Chart as follows. Consider the strengths of the lords of 1 and 8 and of karaka Sani. If all these three are strong then one will have long life extending beyond 66. If two of them good then Middle span and if all 3 are weak short life to be predicted. For longevity both the lord of Lagna and 8 should be strong but the lord of 8 should not be far stronger than the lord of Lagna. A little practice and experience will guide you properly though in the initial stage it may look difficult to estimate. Even in the different spans of Low, Middle and Long life there are lower and upper grades depending upon their strengths which you will also be able to see at a glance from the Rasi Chart. For gauging the strength consider the Shadbala narrated by us The short life is below 32, the medium below 66 and long life extends beyond 66 to 100 and sometimes beyond 100. Gauging thus the span of life from the Rasi chart then look to the maraka working for that span from the position of the dasha lord then acting in D-8. The dasha of a planet badly situated causing maraka and the Bhukti of a similar nature but connected with the Dasha lord will kill the native. (See later lessons for Bhukti readings). Sometimes it may so happen that all the planets in D•8 are situated in good positions other than maraka positions then Sani the karaka takes over the role of maraka and during his dasha he kills the native, Even the classic texts say the same that when there is no maraka planet intervening in age then Sani himself will-take over the role.

While reading D-16.

You may be sure of the reading of this chart only when the Lagna and all the planets or at least the dasha lords fall within the chart but when they fall outside the chart as attaining the Amshas of Brahma, Vishnu, Ishwar and Ravi we cannot be definite on the issue.

But one feature we have noted is that when Lagna or a planet is situated in the Amsha of Vishnu he will have the comforts of conveyance as God Vishnu is the giver of happiness in life. In particular if the Lagna is in Vishnu Amsha all through his life he will be blessed with conveyance and comforts in life irrespective of dashes.

While Reading D-24

While studying this chart you must consider the age of the native and the dashes that run then. As the usual normal ages of study will be between 5 and 25 see the Dasha Lords that rule then and judge the effect. It is not only the higher education of the times (Rajavidya-the education of the era) that may be read out from this chart all other types of study as Vedic, cultural, physical, arts and crafts etc. may also be read out during the dashes of the planets happily posited at all ages. In fact the entire field of education in all branches are to be read out from this chart alone. Whether one attains efficiency in Astrology should be judged from this chart noting the karaka of astrology specially Sani, Sun and Budha being subsidiary.

While Reading D-40.

This chart indicates auspicious and in-auspicious matters such as auspicious functions at home as marriage etc., while in-auspicious ones are the deaths and the consequent obsequies etc.

Note.

Please note that while reading the effects from a D-chart you must confine only to the Dasha and the Bhukti lords of the particular period and none else. Thus D-charts will portray only the periodical changing effects as against the permanent effects revealed in the Rasi chart.

As sufficient number of lives are illustrated in our text from pages 66 to 64 of Pt. II it is not worthwhile to repeat them once again So in this lesson we will only point out some of the chief features to be specially noted while reading a D-chart.

1.   To know how the 'Vargottama position works in a D-chart **see chart-1,** D-9 and D-11.

2.   For Exaltation and Digbala effects see D-10 of Chart-7; 5; 8; 10; 18 and D-24 of Chart No. 1.

3.   For Yogas of Varahamihira see D-10 of Chart-7; D-4 and D-6 of Chart **14; D-4** of Chart-17 and D-10 of Chart-33.

4.   For Widowhood of mother see D-12 of Chart-8.

5.   Generally a good event is likely to take place during the period of Dasha and Bhukti that are situated favorably in the concerned D-chart. This is always for good. But sometimes it may also happen to fall during some badly placed planet's period when it is certain that the event will not end happy. For example the time of happy marriage has to be read out from D-9 during the periods of planets situated happily in the chart. But if it is seen to happen during the periods of planets badly situated such an alliance will not end happily. Either the partner dies or for some reason there will not be marital happiness. So be it remembered that for all happy continuance it is better to wait till the good period comes. This is only human advice but we :know that marriages are made in Heaven and as it should be so it happens. At least in such matters where human discretion may be used this may be followed as in the case of starting a new venture, building of house, fixing moments for doing some acts in life etc. The result of such a mishap is seen in D-9 of Chart-10.

6.   Birth Avayogi in Dustanas in the Rasi chart is

said to cause Yoga but the same principle does not hold good in a D-chart. Under such a position in a D-chart the matter worsens. Vide D-9 of Charts 11 and 14.

7. In the case of Ladies who are mostly Dependents on their husbands when they are under their care (and not when they are under the care of somebody else) see their D-9 for their prospects in life as their prospects depend upon that of the partner in life Vide D-9 Chart 11.

8. In D-1.1 profits and losses will be caused by those that are represented by that Bhava in which the planet is. Suppose the Dasha Lori is in 7 it is from wife and wife side people he gets benefitted, in 9 from father etc., Vide D-11 of Chart-12.

9. Digbala gives a commandeering position in life in matters of the can-notations of the D-chart. In D-10 professionally, in D-11 financially etc., Vide D-10 of Chart-12.

10. Magnitude of effects has to be read out combining the Shadbala strength of the planet both in the Rasi and D-chart. Vide D-10 and D-24 of Chart-14.

11. Association or Aspects with other planets alter the effects. See D-8 **and** D-10 of Chart-16.

12. Budha in 8 in D-4. We have said in aPology with the classics say that all planets in 8 in any D-chart is bad except Budha in D-4. Why ? Classics say that Budha in 8 confers sudden wealth by the death of someone (Mritha Dhanam) or some legacy or hidden treasure coming up suddenly. As D-4 represents wealth chart Budha in 8 in D-4 causes this special good but in other chart he does not do pod. Vide D-4of Chart-7

13.   Dasha plays chief part in the reading of a D-chart. Vide D-10 of Chart-21, D-10 of Chart-19, D-11 of Chart-24, D-9 of Chart-26 and D-11 of Chart-26

14.   Birth Yogi or Duplicate Yogi situated even in ordinary good position in a D-chart causes good with respect to that chart and that too of high magnitude though not endowed with any of the Shadbala. This is the specialty with a Yogi Vide D-10 of Chart-27.

15.   Even a minute's difference in Birth Time causes a lot of difference **in** the readings. See D-1 I of Chart-36.

16.   Nature of Spiritualism has to be read out from D-5.

The type and standard of spiritualism has to be studied from the nature of the planets that cause the yoga and from the Shadbala strength of the concerned planet. One caused by Guru is of the superior type attaining parabrahmatwam (realizing God) as may be seen from Chart-37. One caused by Sani is of philosophical type but a little attached to earth vide Chart-39. In the same chart Budha type is also seen at a later age during Budha Dasha. Budha being a planet with Buddhi and Jnana it causes one of Jnana type. Sukra causes one of showy, glamour, earthy, 'material and magnetic type. Those caused by other planets are not fruitful except that of Sun which elevates one to the Solar world. These various types may be seen from Charts 37 to 40.

Thus we have cited some additional hints here to enable you to follow our new theory propounded by us for the first time in the history of this science which has been highly appreciated all over the world. We hope we have made the

136

subject quite bandy, intelligible, and easily understandable by even a lay man.

So please digest this important lesson which is the crux of our invention and be prepared to receive our next lesson on the more important and illuminative subject—Dasha Bhukti.

LESSON No. 18

Dasha Bhukti Readings

Chapter IV of Pt II of our Text deals with the Dasha Bhukti Readings based on the well-known classic system of Parashara but with slight modifications here and there using Yoga and Avayogi nature of planets, Our classic authors confine entirely to the Rasi Chart which as we have known will not always yield correct and proper results. So we have devised a novel method (New Techniques) where all these classic principles are applied to our Division Charts when you will find astonishing success in all cases without exception.

In lessons 16 and 17 we have dealt with the mode of estimating the merits of a Dasha Lord. In this lesson we deal with the evaluation of the merits of the Bhukti Lord As far as possible we have rendered the subject easy to handle and cent percent success is assured when the planets are singly situated in similar positions from D-Lagna and Dasha lord. When they are with others or aspected by others or in varied positions from the D-Lagna and Dasha Lord some discretion of yours will be necessary using of course the general principles already laid down by us.

Enough is said in our text on this aspect and yet we make it clearer still in this lesson. Broadly speaking you must remember the following dictums to estimate the worth of a Bhukti Lord.

138

1. Unless the Dasha lord be overpowered by the Bhukti Lord it is always the effect of the Dasha lord that prevails. By overpowering is meant that the Dasha Lord be either with or aspected by the more powerful Bhukti Lord or be in the house of or star of the Bhukti Lord,

2. On the other hand if the Dasha Lord be overpowered by the Bhukti Lord then the effect of the Bhukti Lord will come into picture. Here you must remember one thing. If the Bhukti Lord is extraordinarily more powerful than the Dasha Lord then the effect of the Bhukti Lord will rule for that period. If it be ordinary then the effects of both the Dasha and Bhukti Lords will come to pass. It means that in such circumstances the Dasha Lord gets modified.

3. No doubt arises when both the Dasha and Bhukti lords are either both good or both bad when only good or bad will happen respectively. Difficulty arises only when they are of opposite traits-one being good and the other bad so far as the effects of that D-chart is concerned. In such cases we proceed as follows:—

4. A good Dasha Lord who is not overpowered by a bad Bhukti Lord continues to do good only though related (for texts of Relationship please see pages 93 Pt.II). But he may during that period cause a tinge of bad of that Bhukti Lord in addition to Dasha Lord's good effects. If not related only good of the Dasha Lord continues during that time.

5. If the Dasha Lord is overpowered by the bad Bhukti Lord who is also related to the Dasha Lord then only the bad of the Bhukti Lord prevails during the period.

6. Similarly, bad Dasha Lord related to a weaker Bhukti lord who is good will cause the bad effects of the Dasha Lord with a tinge of the good effects of the Bhukti Lord because they are related. But in this case if the bad Dasha Lord be over powered by the good Bhukti Lord then good only prevails. On the other hand if this good Bhukti Lord is not at all related to the bad Dasha Lord the bad of the Dasha Lord only continues.

7. First and foremost we must estimate the character of the Bhukti Lord in that D-chart just as in the case of the Dasha Lord as narrated in our lessons 16 and 17 and then:—

8. Look to their reciprocal positions in that D-chart• Please note that while so doing you must count the position of the Bhukti Lord from the Dasha Lord and not Vice Versa as some do. We are not concerned with the reverse position Viz. that from the Bhukti Lord to the Dasha Lord. Counting this way if the mutual positions be 3-6-8-12 (Dustanas) their bhuktis are generally bad whatever the characters of the Dasha and Bhukti lords be. In such cases some differential readings will arise during such periods. Even though the two lords be happily posited counted from Lagna when they are in such mutually bad positions then some bad will be experienced during that period with all the other good effects that may also be felt during that period. But the finality will not be bad as first importance has to be attached to their positions from Lagna which are of permanent nature while the mutual positions are of temporary nature. If in a D-10 the two lords are placed

140

in good positions from that D-Lagna but in mutual bad positions you may expect some change which one may not like or some mishap in professional matters which latterly proves itself to be good ultimately. If on the other hand the lords are also badly situated from Lagna or at least the Bhukti Lord is so placed then surely that period will be bad. These relative bad positions indicate changes, disruptions, inimical losses, accidents reversions and even death or destruction of the particular effect of the D-chart. On the other hand, if the Bhukti Lord is in good position from the Dasha Lord then you have to read good only provided the Dasha Lord is also good. The best period is when both are in good positions from Lagna and mutually—of course more weight being placed on its position from Lagna.

9. Sometimes the Dasha and Bhukti lords may both be in Dustanas say in 6 and 8 from Lagna and yet may give best of results during their period. This will happen when they form Sankhya Yoga as defined by us in Ch. V Pt. 1• Good will happen if Shubha Sankhya Yoga is formed and bad if Papa Sankhya is formed. So remember if any of the Yogas cited by us exist in the chart. Suppose there is Neecha-Bhanga caused by some planet then their mutual dasha Bhukti will lift the native from low to high position.

10.   C hanges in life will not occur at all ages and so it will not happen in every dasha and Bhukti that changes. It is only at certain specified periods that changes occur and to know them we have to observe the following rules:

a) During the period of the Bhukti of the lord-of the

Rasi or Star in which the Dasha Lord is posited particular effect will be noticed. (This has to be looked into in the Rasi Chart)

b) During the period of the Bhukti Lord who owns the rani occupied by the Dasha Lord in a D-chart (to be seen in D-chart).

c) Planets that are with or aspecting the Dasha Lord. (to be looked into in the Rasi and D-charts chiefly in the D-chart when we want the particular effect of that D-chart).

It is during the period of such Bhukti Lords that definite events happen—the good or bad to be ascertained from the tests enunciated before, as applied to any D-chart we want.

11.   The mutual bad position of the Bhukti Lord from the Dasha Lord in the Rasi Chart no doubt gives indication of some bad to some aspect during that period but what exactly is the bad indicated may not be so easy to assess from the Rasi chart alone.   If you survey all the D-charts for the particular period you will find in some of them such bad positions occurring when you may definitely attribute bad effect pertaining to that D-chart, but not in respect of happily posited D-charts. Likewise a mutually good position existing in the Rasi Chart will speak of good with respects to items connoted by such D-charts in which such a happy position also exists.

12.   We have found out that during the period of a planet situated in own house but in Dustanas neither good nor bad is experienced.   Its effect will at the start appear to be favorable but finally nothing will be realized.   It is for this reason that the

Daivaganas said that even good sign positions in Dustanas are of no avail. For substantial realization of good they should be in good positions from Lagna in a D-chart. Being so, even if their sign positions are bad they still cause good only but with lower degree of magnitude of effects. It merely speaks of the quantum and not nature. We have several times before emphasized that the good or bad is cause by their positions from Lagna while the quantum of effects is based on its Shadbala.

This is how you must look to Dasha Bhukti effects. Is it not a nice, intelligent, definite, logical and easy way of approach – so sure of success that an infant astrologer trained this way can shine far better than the highly renowned one of classic system.

Please go through all the rules enunciated in the texts on this subject as propounded by us and add to them these cornels. Then you will come out with laurels on your head and bring good name not only to you but also to the preceptor and to the very science itself.

Read in the next lesson our masterly treatment of Gochara.

LESSON NO.19.
Gochara / Transits of Planets

In the previous lessons we have understood the art of predicting the general life aspects from the birth Chart and the good or bad effects at particular periods with the aid of Dasha Bhuktis and yet we may sometimes get stranded to gauge the effect at a particular moment as the span of even a Bhukti is too long. So the last recourse left with us is the Transits of Planets-Gochara. This Gochara will also be more or less a ready-reckoner and a surer way of fixing the time of events as we may be sure of the exact position of planet at a particular time. This will also solve the differences in the balance of dasha at birth consequent on the different theories in the astronomical calculations of the position of Moon. Thus, without much trouble and deeper calculations you may give a rough estimate of the times and even the happening of an event indicated by the Dasha and Bhukti lords.

As any other topic is handled in multiple ways this subject of Gochara is so mishandled. The usual way of looking into it from Moon's position or sometimes from Sun's position is on the face of it fallacious for the very reason that millions of births takes place like animals, birds and reptiles are born under the same Moon or sun sign (Rasi) but their lives will not be all alike at any particular time. So it amounts to saying that a such a test may be applied to world predictions rather than to individual lives for which more minute details are to be considered as are particularly applicable to individuals. Thus it

144

amounts to the consideration of one's Lagna and the positions of planets at one's birth as well the dasha and Bhukti running at that time. So Lagna plays the most important part not only for all other purposes but also to Gochara.

Again there is another blunder usually committed which we often come across with in all Newspapers and Magazines which have at late become the attraction of the day which works only to the advantage of the publishers but to the disappointment of the anxious readers. By consideration of all the planets at a particular time what is the definite conclusion that anyone can arrive at. In fact this is not what Daivaganas want you to do. They have definitely said that the effects arising out of a planet has to be judged first from the birth chart, then from its Dasha and finally from its Gochara. So you have to consider the Gochara of the dasha lord only. If you so desire you may also consider the Gochara of the Bhukti lord. This would be more necessary if the dasha lord is a fast moving planet and the Bhukti lord is a slow moving planet like Sani, Guru, Rahu and Ketu. To do this the more sensible way is to reckon it from birth Lagna instead of sun or moon. Even Astakavarga method is a failure in our experience. It is always our research experience that we teach others and not the hypothetical dictums of good old books.

The modes of judgment of Gochara are more fully described in our text Ch. 5, Pt. II. Still we detail below our latest research results for the benefit of our students.

Rules

1.   First note in what bhava reckoned from Lagna the dasha or Bhukti lord is transiting.   By bhava is meant the entire span of a bhava irrespective of its count from lagna-rasi. Roughly, a bhava ranges from about 15^0 behind the cusp of that bhava to about 15^0 above it.   This is only a rough estimate for rough reading while for accurate ones you have to actually work out the spans as per our lessons on Cuspal and bhava charts.   In addition to the bhava it transits note the Rasi it touches, the star in which it moves and its then condition as retrogression, combustion, aspects and conjunctions with other planets.   It is only on consideration of all these factors that you will be able to diagnose its proper value and not by the superficial survey of its position from radix moon.

2. Considering the above factors the mode of judgment is as follows:-

a.   If the dasha lord is good at birth (note that birth time qualities are most effective in shaping the final effect then he gives out good results while transiting good bhavas, Rasis and stars.   Of them the bhava position is of paramount importance.   For example, one transiting even in exaltation sign may not give good effects when the bhava transiting is 8th.   When this good lord transits unfavorable positions then by virtue of his birth right goodness he will not cause any harm though he may not be able to give much good.

b.   A dasha lord who is bad at birth transiting any position except his own house gives out his inherent bad effects only and when he moves in unfavorable positions causes immense bad.

c.   It then follows that a mixed lord will give mixed effects – good in good positions and bad in bad positions. Thus you see that the birth character is paramount importance.   Are these factors considered by the present day astrologer.   Then how can those forecasts be relied upon.   Hereafter never even glance at such dubbings.

d)   A good lord by birth qualities getting retrograde in transit gives good effects while a bad lord being so causes worse results.

e)   Good or bad lord passing through Zero Rasi or getting into combustion in transit does not do good in whatever bhava he may move. But, Rahu, Ketu and Birth Avayogi transiting Zero Rasis will do well provided they transit good bhavas

f)   Retrogression in a Zero Rasi while in transit attains strength and its effects are to be predicted on the above lines considering the good or bad nature at birth

g)   In the case of slow moving planets who move

out of a sign after a long time it may be difficult to estimate as to when exactly it may give out its effects in that sign. So we have to look to the star in which it is passing. The results arising' from this will have to be read out as we once did in the case of planets situated in particular stars.

h) Next look to the conjunctions and aspects of other planets on this dasha or Bhukti lord whose Gochara you are looking at. The lords of Dustanas, the enemies of the dasha lord and deadly Malefics as Ketu, Rahu, Sani and Kuja aspecting or conjoining the lord will hamper the good while those of benefits do good. This aspecting or conjunctive effect lasts till 3' orb or range. It is forcing while applying and fading while separating. For applying and separating aspects see our lessons on aspects.

i) One definite thing that is revealed as astounding truth is that when the dasha lord transits the actual !birth yoga star the native will catch a new and fresh order of prosperity in life. In the same way when he transits the actual Avayogi star he causes bad. The actual Avayogi star is about 180' from the yoga star. It is not exactly 180' but nearby it belonging to birth Avayogi. As the 6th dasha lord from Birth Yogi becomes Birth Avayogi it is the 15th star from birth yoga star that becomes birth Avayogi star. For example if the yoga point falls at 29' Mesha belonging to birth yogi Sun it is not the exact 180th degree point from this i.e., 290 Thule where Vishakha star is situated that becomes Avayogi star but Anuradha the one next to it as birth Avayogi of Sun is Sani. Further all events in life invariably happen on the Thithis, Weekdays, and stars of the Dasha or the Bhukti lords.

      j)   Even after consideration of all the above factors we may not be able to satisfy the wants of our consultants as to the exact or more probable times of happening of events due to very slow or very fast movements of planets. So we suggest to you to make use of the Western theory of Progression and among them we prefer to value the progressions of Moon and Tenth Meridian only. The former indicates all aspects in life while the latter points out only professional aspects. A word about this Progression. A days position from birth date and time of birth will be the progressed positions for a year and on this basis Progressed Moon's annual motion may be taken as about 12' a year for all rough purposes and at I' a year for Progressed Meridian. Of course the actual position will have to be worked out from the Ephemeres for the number of days past as years past and for the same time as that at birth. Then see in what bhava this Progressed Moon is moving and deduce the results as in the case of transit of dasha lord we have already discussed. The probable time of occurrence of an event will be when it forms aspects with the Radix positions of planets and cusps (positions as at birth time). Note the time when this progressed Moon touches the birth yoga star when some fresh yoga starts. Likewise when it touches Avayogi star some bad is seen· It causes good while passing through good bhavas and bad otherwise (as usual bhava spans to be noted here also) In the case of Progressed Meridian its movement in bhava has no significance. See if it forms any good aspect with the radix positions of planets, especially the dasha and Bhukti lords and birth yogi and Avayogi planets. Passing through the actual birth yoga or avayoga stars will also cause effects—good or bad respectively. The good aspects are according to western astrology 30', 60', 120' and sometimes conjunction· We find that even opposition does good when those planets are yoga planets at birth (functional benefits). Westerners ascribe all timings whenever there be aspects with any planet but in practice we see that it is more true when such aspects are with the Dash& or the Bhukti lord. If per chance this Progressed Meridian were to form fine aspects or conjunction with radix Birth Yogi or its actual yoga star then as certain as anything it shall be a special period in his life to catch good luck in profession, honor and status.

In the chapter on Gochara we have in our text dealt with various other topics which are more or less of doubtful character and full of alternatives. As these do not help our students at this stage we have included them under our Visharada course as we mainly confine to definite dictums under this Pundit Course since our aim is not to confuse and confound the students with all sorts of unworkable theories. See our next lessons.

Lesson No. 20

I n this lesson we clarify some of the misconceived subjects that are very strongly rooted in the minds of many persons while in reality they are wholly wrong.

1. Rahu kalam.

It is usually calculated as lasting for 11/2 hours in some fixed part of a weekday as published in all Panchangas, Diaries and Papers. This is not quite correct. It has no doubt a fixed part in a weekday (Vide PP. 156 Pt. II)—say for example on Friday the 4th part and so on. But its duration is not always 11/2 hours. To get the correct duration you must divide the duration of Day (from sunrise to sunset) into 8 equal parts and commence the first part from sunrise of the place on that day. The duration corresponding to that part of the day fixed for the weekday will be the actual duration of Rahu kalam. This is the correct mode.

The general conception of Rahu kalam is that it is always had and has therefore to be avoided for all auspicious undertakings. Yes, you may avoid it for general public functions as the opening ceremonies of some public institutions, of organizations, etc., where individual horoscopes do not play part. But, in the case of individuals where one's own luck plays part his horoscope has to be scrutinized and this Rahu kalam judged. If Rahu at birth is yoga karaka and when Rahu Dasha itself operates you will see that all good things will happen during Rahu Kalam alone whether you like it or not. In such cases you may definitely advise persons to take to undertakings in Rahu Kalam alone. (Of course other than ceremonial functions as marriage etc.). If Rahu is badly placed at birth you have to avoid Rahu Kalam in lifetime. Thus you see the good or bad of Rahu Kalam depends chiefly on the merits of Rahu at birth time and not merely by his deadly name.

2. Thithi.

Under the consolidated list of Karakatwas of Planets in our Part I we have mentioned the Thithis ruled   by planets. For the benefit of our students we stats below the list.

Planets.	Thithis they rule.	No of Thithis.
Sun	Prathama-Ekadashi	1 - 1 1
Moon	Dwithiya-Dwadashi	2 - 12
Kuja	Shasti	6
Budha	Sapthami	7

Guru	Astami-Thrayodashi	8-13
Sukra	Navami-Chaturdasi	9-14
Sani	Dashami-Poornima	
	& Amavasya	10 – 15 – 30

You know that planets give their effects during their Dashas, bhuktis, and months, Thithis, Weekday, Star and Yoga. For example if Sani be good in the horoscope and his dasha rules then be will give his good effects on Dashami or Full-moon or New moon day as they are controlled by Sani. Similarly if he bad at birth the bad effects are felt on those days. If suppose good is experienced on a new moon day one may not wonder. This will be more effectively felt if sun and moon who cause new moon are also well placed at birth and in transit this new moon day falls in good bhavas counted from birth Lagna. By this is meant that sun and moon should both be together in auspicious bhavas. For example—for one born in Vrischika Lagna Sun is lord of 10 and Moon lord of 9 and their conjunction is excellent as lords of 9 and 10· If during transit they combine together in any auspicious Bhava say in Vrishbha the 7th bhava then that new moon day proves good as it falls in 7th bhava (please note that in reckoning the bhava the actual span of the bhava should be considered), The same

154

may be experienced during the period of the dasha of Sun or Moon. In this way you must judge the effect of a Thithi by the planet that controls the Thithi and not by the superficial and the more general way that are applicable to world. Thus, you realize that for a native every Thithi is good or bad depending on the merits of the planet controlling that Thithi.

4. Weekday.

Generally Tuesday and Saturday are treated as being bad days. Why? Because their lords Kuja and Sani are termed as deadly natural Malefics. This may be so for the world and certainly not to the individuals. The good or bad of a weekday depends upon the good or bad of the planet that controls the weekday. If in the above case Kuja and Sani become functional benefits attaining all positive characters then those weekdays will be good especially during their Dashas. On the other hand the so-called good days of Guru and Sukra (Thursday and Friday) may prove fateful if they are badly situated in birth chart and also their Dashas run at that time This is how you must look to the good or bad aspect of a weekday.

5. Nakshatra (Star).

Volumes and volumes are said on the bad effects of certain stars (Nakshatra Dosham) and many books are published on same but none of them fit into actuality - So, be not misguided by such unauthentic and impracticable sayings. In fact a star is good or bad according to the merits of the lord of that star i$_n$ the horoscope In addition, you have to reckon its position from Lagna. A star situated in Dustanas is not good though it be benefit otherwise. For example in the case of Dhanur Lagna Ashlesha star situated in 8th bhava is bad but do not suddenly jump into any conclusion on this one count Further see where its lord Budha is If he be in good bhavas say **in** Mithuna the 7th bhava then the bad effect of the star Ashlesha being in 8 gets modified. If on the other hand he be in 3 or 8 then predict death, d'smissal, degradation, etc., during the period of Budha especially when he transits Ashlesha star.

In this way the good or bad of a Thithi, Weekday, Star **or** Yoga should be judged considering their merits in the horoscope and not by merely valuing by its general characters. As we have mentioned about Yoga too we shall speak a word of it now.

6. Yoga.

In general astrology some of the Yogas like Vyathipatha and Vaidriti are spoken of as very bad ones. This is again t^{h}e fallacious way of valuing a yoga merely by its general characters. The real good or bad of any yoga with respect to an individual has to be judged by its lord (the lord is the stellar lord of the star in which the yoga is situated) (see Pt. I of our text). For example the

156

lord of Vyathipatha yoga is Rahu as this yoga falls in the star Shatabhisha whose stelᒿlar lord is Rahu. If Rahu is well situated at birth then this yoga becomes good to him even though it is said to be generally bad. In this way we have to conclude these merits by considering their intrinsic value existing in the Birth Chart and not by the general sayings.

7. Kalasarpa Dosha,

This is another misconceived subject. This happens only when all the seven planets are in between Rahu and Ketu and that in a particular order. They must be in the zone counted from Ketu to Rahu in clockwise order or in other words in anticlockwise order counting from Rahu to Ketu. But, if the seven planets are in the contrary part of the zodiac then there is no dosha under Kalasarpa. Many do not know this distinction. Even here care must be taken to see in the case of planets that are in conjunction with Rahu or Ketu whether they fall within the above defined zone or not reckoning the degree positions of the planets in conjunction and that of Rahu or Ketu with whom the planets conjoin. Suppose Rahu is in Mithuna and Ketu in Dhanush and all the six planets other than sun are in between Makara to Vrishbha and Sun is at 4' Dhanush while Ketu is at 15' Dhanush. Here all the six planets are no doubt in the portion from Ketu to Rahu in clockwise order. As Sun is in the other zone not falling in between Ketu to Rahu in clockwise order this one planet Sun breaks the entire formation of this Kalasarpa Yoga. If on the other hand this Sun was at 20' Dhanush then this Yoga would have been formed. In this way this subtle difference has to be noted. Again the bad effect of this Yoga will be caused only during the period of Rahu or Ketu Dacha or during the dacha of a planet situated in their stars. Its effect is to smash down all the good enjoyed by the native so fug the later part of life being rendered miserable.

If Rahu at birth be birth yogi under our theory or even a yoga karaka under the classical theory you should not read the bad effects of this yoga. The Bhava in which Rahu is situated in such a yoga plays a very important part. if in reality there be kalasarpayoga then that bad commences with the effects of the bhava in which Rahu is. For example if the said yoga be formed when Rahu is in 9 father or Paternal wealth will cause ruination, if 5th the son, if 7th the wife and so on. In all the above arguments we centered round Rahu, only because this yoga commences from Rahu and ends with Ketu. This is how you must estimate Kalasarpa Yoga.

7. Lastly Kuja Dosha.

Be not led away by the threats of the old school of astrologers who by the mere existence of Kuja in 1 2 4 7 8 12th places from Lagna, Moon or Sukra ascribe widowhood or widower hood. At this rate almost all horoscopes have this dosha in one form or other. For fuller discussions on this subject please read PP. 162 PT. I. In fact all Malefics cause this widow hood and why should Kuja alone be singled out – may be for the reason that he is 'Mangala Karaka'. Unless you weigh the true color of Kuja in a horoscope you should not offhandedly accuse him and thus bear his curs

To judge the aspect of Mangalyam (Womanhood) you must weigh carefully the merit-of the 7th and 8th bhavas and also the condition of Sukra who is karaka of the partner and the dashes that operate and finally conclude your opinion. It is by such deep investigation only you can judge this aspect and not by the superficial positions of Kuja. 1n this connection we wish to bring home to you a fallacious argument that the existence of Kuja dosha in both the horoscopes (husband and wife) will have a cancellative effect which is absurd. Pure logic and commonsense tells you that the dosha in both will simultaneously harm both. A more sensible way of approach to this problem is to judge the longevity of the other when this dosha is seen in one.

Thus we have fully apprised you of the fantastic notions of many misconceived topics and we hope that at least in future you will not fall a pray to such generalized sayings and also hope the public will grow wiser. Read our next lesson on Marriage Alliance Tests which will be very interesting.

LESSON No. 21.

Marriage Alliance

In this lesson we treat the subject of Marriage Alliance Tests which is of great importance as you may be often consulted by your friends or clients. This is again mishandled by most of the present day astrologers. In fact, it is mostly disposed of by the family Purohit who does not know even the rudiments of astrology, He follows the old theory of Koottams based on the birth stars of the two which are published in all Panchangas and cited in Kalamrit text, By this one test they come to a very candid and definite judgment as if astrology is pure mathematics. If they happen to see more than 18 marks they second the matching. Thus these Koottams are the sole guide to them while in reality they are after all very superficial cannons. The real line of perspective is to judge the intrinsic worth of a horoscope. What the old school of astrologers do is merely the superficial survey based on the birth stars of the two. It is not only the birth stars that play the part here but you should also consider the birth lagans and the planetary positions as a whole. What is roughly estimated by way of Koottams may be more accurately evaluated from the body of the horoscopes concerned. This method is surer than the other.

How are you sure about the accuracy of the other horoscope when you cannot bank on the accuracy of your

160

own, it is absolutely necessary to go into both the horoscopes? It is not possible to judge the fate from a single horoscope only. Do you not believe that marriages are made in Heaven. Then does it not mean that one's fate is already sealed. Thus we have to estimate all about the partner's affairs and marital happiness from a single horoscope alone. Then how to do it. Please stud; carefully pages 162 to 165 Pt. II of our text as all these aspects are fully discussed there.

Instead of trying to find out the feasibility of alliance between the two horoscopes if you can guess the destiny between them you would be rendering a higher service and relief to your friend who has already labored heavily in the search of several horoscopes which might have failed with him for one reason or other. In addition to the tests of destiny described in our text see if the effects of the dashes that run thereafter in both the cases are alike. If the husband has to have good times ahead then the coming dashes of the wife should also be good, Vice Versa. This can be judged better from D-9 of the other than D-1. Likewise see if the marriage time has come in both. If one indicates the approach of time while the other not then conclude that they are not destined match. These should be judged by dacha Bhukti and Gochara as taught in the previous lessons.

Please note here that as marriages are made in Heaven the destined coupling can never be averted by human trials as the hand of God works topmost even if one is cautioned against coming evil. All these are the outcome of Fate which cannot be overcome by common men like us who are attached to earth. There should we look into the other horoscope at all. Yes, you may not, is our candid reply. For instance, -if one has strong indication of widowhood in early age she will certainly join one of abort longevity, vice versa. That by marrying a particular girl one's luck gets enhanced or lowered is a tall and empty say. In fact no other person can mend or end the fate of one. What really happens is-the two events synchronize. In his horoscope indications may be that the marriage and prosperity start at the same time. The planet that rules at that time will be good both in D-9 and D-10 or D-II. Likewise in some cases every birth of a child brings on good luck to the father. In this case the planet or planets that rule at those times will be good both in D-7 and D-!O or D-11. This coincidence of events which is already existing in the horoscope is interpreted as coming from the other. Let those who believe that the wife is a Lakshmi to him go so. In a way it is good. Poor wives will get better treatment at the hand of the husbands. But let them not dislike or hate if anything bad happens after the event.

Survey of a single horoscope alone will do to arrive at proper judgment, In the Rasi chart see how the dasha lords that rule later are disposed with respect to 7th and 8th bhavas and also see how the lords are disposed in D-9 and then conclude. By this mode of judgment you will be nearer the mark and far better in judgment than the crude old method of Koottams. Lastly, for consolation sake you may look into Koottams and among them Nadi and Rajju may be preferred. Even if these two are wanting you may go ahead if the intrinsic worth of the horoscopes are good enough so far as matrimonial affairs are concerned. For verification with Koottams you need not labor as most of the Panchangas of the day give them and some give in ready-made tabulated form which gives out the result at a glance.

Lastly we take up Muhurtams (Election Time).

Texts on Muhurtams have levied certain hard and fast rules to be observed in the fixture of auspicious times for the performance of certain acts and if you are interested you may go through any book on Election Astrology and one of them is Kalamritam a Sanskrit text available in all languages. If you go through them you will see that they are after all of the nature of a far general character applicable to a mass than one pointing to any particular individual. In reality what is good to one may not be so to the other like the proverb 'what is food for one may be poison to the other'. What is really needed is the fixture of time suited to an individual. To do this you must look into one's horoscope and then decide what day and time is propitious to him. In fact, it can be said that there are both good and bad times on all days which depends upon the selection of the moment. Still we have to respect and revere the broad principles especially in the case of major auspicious functions as Marriages etc. For other minor matters you may select any day that proves to be good in consonance with the particular chart at birth. The chief factors to look into are the birth Lagna and the dasha that runs at the time. These should be in happy .positions from the Muhurta Lagna. All look into 7th and 8th positions from Muhurta Lagna and then decide. This is not enough. You must also see how the dasha lord of the Muhurta time is posited in Muhurta chart. So long as the current dasha and the succeeding ones are good this matter runs well, but it goes bad when this aspect also goes bad. The Muhurta chart is nothing but the birth chart of the particular aspect considered. The particular bhava and the karaka of the event should be scrutinized in the Muhurta chart and results read out coupling of course the current dasha and that after. For instance in the case of marriage the 7th house and Sukra, in the case of house construction 4th house and Sukra and Kuja (Kuja is Bhoomikaraka and Sukra is Grihiha karaka) etc., should be scrutinized. In this way you should weigh the pros and cons of Muhurta and not in the most generalized aid abstract way of the old school. For further details please read pages 165 to 167 Pt. 11 of our text

With this we have finished Part II of our text and have completed all the subjects under our 'Pundit Course'. There are other topics discussed in our text which are after all of academic interest. One interested in them may study himself They are reserved for higher course of 'Visharada'.

We hope you have by this time digested our theories well and are in a position to stand a test which will be conducted by us in our later lessons. Though titled as question papers they are mostly of the capitulatory character which is more or less revision lessons. You will have to send answers to those questions. After you satisfy us we will issue the necessary certificate. If there is any balance of fees to be paid please remit the same so that all the 9 question papers may be sent at a time.

Please note that we have the noble idea of spreading our system far and 'wide. Our final appeal to you is that if you intend rendering any real service to your Guru and to this science you should spread this theory far and wide and see that like you some of your interested friends get the benefit of this College.

Finally, we bless you whole-heartedly to have a masterly knowledge of this science and try to even outshine your Guru.

MAY GOD BLESS YOU

ABOUT THE AUTHOR

Author is Bachelor of Science, Bachelor of Mechanical Engineering, and Master of Food Technology (Dipl-Ing) from Germany. I never believed in Astrology till I met my Guru Mr. Sheshadiri Iyer from Karnataka, India. He predicted that I shall learn Astrology from him and develop this science for the welfare of the humanity. I have added these predictions and technique for prediction for the benefit of my readers.